UNDERSTANDING THE IMPACT OF HUMAN INTERVENTIONS ON THE HYDROLOGY OF NILE BASIN HEADWATERS, THE CASE OF UPPER TEKEZE CATCHMENTS

Tesfay Gebretsadkan Gebremicael

UNDERSTANDING THE IMPACT OF HUMAN INTERVENTIONS ON THE HYDROLOGY
OF NILE BASIN HEADWATERS, THE CASE OF UPPER TEKEZE CATCHMENTS

DISSERTATION

Submitted in fulfillment of the requirements of

the Board for Doctorates of Delft University of Technology

and

of the Academic Board of IHE Delft Institute for Water Education

for

the Degree of DOCTOR

to be defended in public on

Tuesday, 8 October 2019, 15:00 hours

in Delft, the Netherlands

by

Tesfay Gebretsadkan GEBREMICAEL

Master of Science in Water Resources Management

IHE Delft Institute for Water Education, Delft, The Netherlands

Born in Adwa, Ethiopia

This dissertation has been approved by the promotor and copromotor

Composition of the Doctoral Committee:

Rector Magnificus TUDelft	Chairman
Rector IHE Delft	Vice-Chairman
Prof. dr. ir. P. van der Zaag	IHE Delft and Delft University of Technology, promotor
Dr. Y.A. Mohamed	IHE Delft, copromotor

Independent members:

Prof. dr. ir. H.H.G. Savenije	Delft University of Technology
Prof. dr. W. Bewket Alemayehu	Addis Ababa University, Ethiopia
Prof. dr. G.P.W. Jewitt	IHE Delft and Delft University of Technology
Prof. dr. ir. R. Uijlenhoet	Wageningen University & Research
Prof. dr. ir. N.C. van de Giesen	Delft University of Technology, reserve member

This research was conducted under the auspices of the Graduate School for Socio-Economic and Natural Sciences of the Environment (SENSE)

CRC Press/Balkema is an imprint of the Taylor & Francis Group, an informa business

Published by:
CRC Press/Balkema
Schipholweg 107C, 2316 XC, Leiden, The Netherlands
Pub.NL@taylorandfrancis.com
www.crcpress.com – www.taylorandfrancis.com
ISBN: 978-0-367-42508-1

ACKNOWLEDGEMENTS

Writing this thesis has been fascinating and extremely rewarding. First and for most, I would like to thank God for his protection and giving me a strength to complete this work. Without his blessings, this achievement would not have been possible. I would like to thank all who in one way or another contributed in the completion of this thesis.

I am gratefully to acknowledge to my esteemed promoter Prof. dr. ir. Pieter van der Zaag for giving me the chance to study under his supervision. I greatly appreciate all the scientific guidance starting from my master thesis to the four years PhD study. I respectfully thank you for the trust, the insightful discussion, offering valuable advice, for your support during the whole period of the study, and especially for the timely response of my e-mails, critical comments, corrections and suggestions during writing articles. I will not forget those entertaining and helpful discussions with you at each step of my study. Without your continuous encouragements and supervision, I would not have completed this thesis.

I have great pleasure in acknowledging my gratitude to my co-promoter Dr. Yasir Abbas Mohamed for his regular guidance, critical comments, corrections, suggestions and reviewing and shaping of the thesis to its current form. From Dr. Yasir, I learnt, not only the vast experience and knowledge in this field of catchment hydrology, but also how to write scientific paper in short and precise way. Without your encouragement and constant guidance, I would not have finished this thesis within four years. I also acknowledge my local supervisor Dr. Eyasu Yazew for his guidance, sharing ideas and the fun that made life easier during my field work in Ethiopia.

I am sincerely grateful to the Netherlands Fellowships Programme (NFP) for their financial support in the form of a scholarship to undertake this study. I am also so thankful to the Tigray Agricultural Research Institute for the financial support and providing me vehicles during my field work. I would like also thank Ms. Jolanda Boots, PhD fellowship officer at IHE Delft for her continuous support in arranging air tickets during traveling for field work and overall handling financial issues during my study. I gratefully acknowledge data and information provided by the Ethiopian Meteorological Agency, and the Ethiopian Ministry of Water Resources, Irrigation and Electricity.

I also thank my family and friends who encouraged me, supporting and allowing me to be away for part of my study. I would most thank my wife Tirhas Abraha for all incredible sacrifice during my first year of the PhD journey. You were alone during your pregnancy of our daughter, Heran. I will not forget what you've made on my behalf during your pregnancy and delivering of our daughter. No matter what, I will love you forever, Titiye; no matter how much we argue, or how much I upset you, I'll love you unconditionally till the day I die. My son Nathan and my daughter Heran also deserve my appreciation for your patience and grief while both your mother and I were away. Your sacrifice for eight months without mother and father will be with me like a handprint on my heart. Last, but not least I am deeply grateful to my mother, Kidan

Welderufael, and my father, Gebretsadkan Gebremicael, who have prayed for me from the first day of my life.

Tesfay Gebretsadkan Gebremicael

Delft, The Netherlands

October, 2019

SUMMARY

The Tekeze-Atbara river basin, one of the largest tributaries of the Nile basin, and shared between Ethiopia and Sudan, is crucial for economic development and environmental sustainability in the region. The upstream part of this river basin, and in particular its headwaters comprising the Upper Tekeze basin, is currently the focus of the Ethiopian government for economic development in the semi-arid areas of Northern Ethiopia. The government has committed itself to an ambitious plan to the eradicate food deficit of the region by utilizing water resources of the basin for irrigation and hydropower developments. Nevertheless, land degradation, water scarcity, and inefficient utilization of the available water resources are bottlenecks for achieving this ambitious goal. Availability of water resources for economic development in the region has been influenced by various natural and anthropogenic factors. Therefore, understanding variability and drivers for change of the hydrology of the Upper Tekeze basin and its implications on water availability is vital for enhanced water resource management in the region.

The Upper Tekeze basin ($\sim$45,000 km^2) is characterized not only by severe land degradation and moisture stress but is also known for its recent successful integrated catchment management experience. It is claimed that catchment interventions have caused enhanced water availability at different locations in the upper parts of the basin. But it is not scientifically known how these human-induced environmental changes affect the hydrological processes and what the downstream implications are. Previous studies focused at either experimental plots or very small watersheds, from which it is difficult to extrapolate and infer basin-wide implications. Therefore, this study aims at better understanding the impact of anthropogenic catchment management dynamics on the hydrological processes at different spatio-temporal scales and their implication to downstream flows. This is achieved by combined use of primary and secondary information, remote sensing data, laboratory analysis and assessment using hydrological models.

Satellite rainfall products are an important source of rainfall information in data scarce catchments such as the Upper Tekeze basin. First, the performance of eight widely used satellite-based rainfall estimates (TRMM, CHIRPS, RFEv2, ARC2, PERSIANN, GPCP, CMap and CMorph), were evaluated against 34 ground measurements to identify which products are suitable for the Upper Tekeze basin. Comparison of these products with the observed rainfall were conducted using different statistical indices at different spatial and temporal domains. The result shows that the rainfall data of CHIRPS outperformed all other products at all temporal and spatial scales. Next, estimates from RFEv2, 3B42v7, and PERSIANN products were closest to measurements at rain gauges for the daily, monthly and seasonal time scales, and both at point and spatial scales. The performance of all products improved as the temporal scale increased to monthly and seasonal time step at all spatial scales. Compared to low altitudes, the Percentage Bias (PBias) at high altitude increased by 35% whilst the correlation coefficient (r) dropped by 28%. CHIRPS and 3B42v7 products showed best agreement in mountainous terrains. CMorph and 3B42v7 consistently overestimated rainfall relative to all rain gauges during the pixel-to-point rainfall comparison approach and at lowland areas during the areal averaged rainfall comparison. The other six products underestimated rainfall at all spatial scales. In summary, rainfall estimates by CHIRPS have the best agreement with ground observations of rainfall in all conditions. CHIRPS was there fore used for validation and filling in of missing ground rainfall data and as an input for hydrological modelling in this study.

The long-term trends and linkages between rainfall and streamflow were analyzed for 21 rainfall and 9 streamflow stations to identify the possible drivers of changes of streamflow in the basin. Trend analysis and change point detection of these variables were analysed using Mann–Kendall and Pettitt tests. Fluctuations in these variables were also investigated using Indicators of Hydrological Alteration (IHA). The trend and change point analysis found that 20 of the tested 21 rainfall stations did not show statistically significant changes during the last 63 years. In contrast, the streamflow showed both significant increasing and decreasing patterns. Six out of the nine streamflow gauging stations showed a decreasing trend in the dry (October to February), short (March to May), main rainy seasons (June to September) and annual totals. Only one out of nine gauging stations experienced a significant increase in streamflow during the dry and short rainy seasons, which was attributed to the construction of Tekeze hydropower dam located upstream of this station in 2009.

The streamflow trends and change point timings were found to be inconsistent among the stations. Changes in streamflow without significant change in rainfall suggest other factors than rainfall drive the change. This indicates that the observed changes in streamflow regime could possibly be attributed to the change in catchment characteristics.

Therefore, first the spatiotemporal changes of Land Use/Cover (LULC) were analyzed. To this effect, the human-induced landscape transformations in the Geba catchment (~5,000 km^2), headwaters of the study basin, were investigated for the last four decades (1972-2014). Satellite images, Geographic Information System (GIS) and ground information were used to classify and change detection of the LULC of the catchment. Furthermore, a probability matrix identified systematic transitions among the different LULC categories and showed that more than 72% of the landscape has changed its category during the past 43 years.

Natural vegetation cover reduced drastically with the rapid expansion of agricultural and bare areas during the first two decades. However, it started to recover since the 1990s, when some of the agricultural and barelands turned into vegetated areas. Natural forest land showed a continuous decreasing pattern until 2001, after which it increased by 28% in the last period (2001-2014). The increase of vegetation cover is a result of intensive watershed management programs during the last two decades. These findings were important for improving our understanding of the relationship between hydrological processes and environmental changes in the basin.

This relationship was investigated using an integrated approach composed of hydrological response of LULC changes, assessing the alteration of streamflow using Indicators of Hydrological Alteration (IHA) and quantifying the contribution of individual LULC types to the hydrology using Partial Least Square Regression model (PLSR). A spatially distributed hydrological model based on the Wflow-PCRaster/Python modelling framework was developed to simulate the hydrological processes of the LULC change patterns previously identified. The results show that the expansion of agricultural and grazing land at the expense of natural vegetation increased surface runoff by 77% and decreased dry season flow by 30% in the 1990s compared to 1970s. However, natural vegetation started to recover from the late 1990s and dry season flows increased by 16%, whilst surface runoff and annual flows declined by 19% and 43 %, respectively.

Pronounced variations in changes of streamflow were noticed in Siluh, Illala, Genfel sub-catchments, mainly associated with the uneven spatial distribution of land degradation and rehabilitation. However, the rate of increase of low-flow halted in the 2010s, most probably due to an increase of water withdrawals for irrigation. Changes in hydrological alteration parameters were in agreement with the observed LULC change. The PLSR analysis revealed that most LULC types showed a strong association with all hydrological components. These findings demonstrate that changes in hydrological conditions may indeed be attributed to the observed LULC change dynamics.

Furthermore the impact of physical soil and water conservation interventions on the low flows were quantified using a combination of paired (control and treated) and model-to-model ("before and after" interventions) comparison approaches. The overall impact of LULC change cannot uniquely identify the quantified impact of such interventions. Large scale implementation of physical Soil and Water Conservation (SWC) structures can modify the hydrology of a catchment by changing the partitioning of the incoming rainfall on the land surface both in positive and negative ways. Therefore, a scientific understanding of the response of low-flows to SWC interventions is critical for effective water management policy interventions.

Results revealed that the fully treated sub-catchment ($\sim$500 km^2) has experienced a significant change in the low-flow response following intensive SWC implementation in large parts of this sub-catchment. Compared to the control catchment, low-flow in the treated catchment was larger by more than 30% whilst the peak direct runoff was lower by more than 120%. This could be explained by that a large proportion of the rainfall in the treated catchment infiltrates and recharges groundwater which later contributes to streamflow during the dry seasons. The proportion of soil storage was more than double compared to the control catchment due to the SWC interventions that improved infiltration capacity of the catchment. Hydrological comparison in a single catchment (model-to-model) also showed a drastic reduction in direct runoff (>84%) and an increase in low flow by more than 55% after the SWC works. These findings were confirmed by the observed changes in hydrological regime using the IHA method. However, whereas the low-flow in the catchment significantly increased, the total streamflow declined significantly after the large scale SWC implementation, which is attributed to the increase in soil moisture storage and evapotranspiration, including from irrigated plots that withdraw water from streams. This negatively impacts the availability of blue water resources to downstream users.

In conclusion, this thesis has shown that the ongoing watershed management interventions in the Upper Tekeze basin have impacted water availability differently at different spatial and temporal scales. Significant changes in the magnitude of streamflow components (e.g., annual totals, wet and dry season flows) were found at all spatial scales. However, the rate of change in streamflow over larger catchments appear to be smaller compared to the smaller catchments. This typical scale effect is mainly associated with the uneven spatial distribution of watershed management interventions in sub-catchments.

Based on the results obtained from the long term trend analysis of rainfall and streamflow, detailed long term land use/cover change analysis, modelling of hydrological response to land use change and the quantification of hydrological response to SWC interventions, this thesis has improved the understanding of how human-induced environmental changes impact on hydrological processes of the Upper Tekeze basin and has been able to quantify their impacts. The combined analysis of rainfall-runoff modelling, alteration indicators and PLSR is recommended to assess the impact of environmental change on the hydrology of complex catchments. The IHA method is a robust tool to assess the magnitude of streamflow alterations whilst the PLSR method can identify which LULC is responsible for this alteration. The results of this study may guide informed water resources and catchment management at different scales, both upstream and downstream of the Upper Tekeze basin, in such a way that the basin may develop in a sustainable manner.

CONTENTS

LIST OF SYMBOLS

Symbol	Descriptions	Dimension
R^2	Coefficient of determination	(-)
r	Coefficient of correlation	(-)
PET	Potential Evapotranspiration	$(L.T^{-1})$
AET	Actual evapotranspiration	$(L.T^{-1})$
P	Precipitation	$(L.T^{-1})$
N	Numbers of pairs of product	(-)
CV	Coefficient of variation	(%)
S	Man Kendall statistics	(-)
n	Length of datasets	(-)
Z	Man Kendall's normalized statistics	(-)
$V(s)$	Kendall's variance	(%)
Z_{cr}	Critical value of Z	(-)
Y_I	Trended series for time interval t	(*)
Y_t	Datasets after auto-regressive	(*)
P	Probability	(-)
T_g	Total change	(%)
N_g	Net change	(%)
L_p	Loss to persistence ratio	(-)
G_p	Gain to persistence ratio	(-)
P'	Amount of water needed to saturate canopy	$(L.T^{-1})$
$\bar{R}$	average precipitation intensity on saturated canopy	$(L.T^{-1})$
E_w	Evaporation from canopy	$(L.T^{-1})$
P_t	Proportion of rain diverted to streamflow	$(L.T^{-1})$
θ_s	Saturated soil water content	$(L.T^{-1})$
θ_s	Residual soil water content	$(L.T^{-1})$
U_s	storage	(L)
U_d	Deficit	(L)
K_{sat}	Saturated hydraulic conductivity	$(L.T^{-1})$
st	Saturated store	$(L.T^{-1})$
K_o	Saturated hydraulic conductivity at soil surface	$(L.T^{-1})$
f	Scaling parameter	(L^{-1})
M	Model parameter determine the decrease in K_{sat} with soil depth	(L)
β	Element for slope angle	(Degree)
WR	Wet root number	(-)
SN	Sharpness parameter	(-)
CS	Scaling parameter to multiply the potential capillary rise	(-)
CSF	Model parameter	(-)
N	Manning N parameter for the kinematic wave function	$(L^{-1/3}T)$

LIST OF ACRONYMS

CHIRPS	Climate hazard Group Infrared Precipitation with Stations
DEM	Digital Elevation Model
DLR	German Aerospace Centre
EROS	Earth Observation Service
EGU	European Geoscience Union
FAO	Food and Agriculture Organization
FLEX-TOPO	Topography driven conceptual modelling
GIS	Geographical Information System
GIUH	Geomorphological Instantaneous Unit Hydrograph
GLUE	Generalized Likelihood Uncertainty Estimation
GPS	Global positioning system
GTP	Growth and Transformation Plan
IPCC	Intergovernmental Panel on Climate Change
ITCZ	Inter Tropical Convergence Zone
IAHR	International Association for Hydro-environment and Research
ISODA-TA	Iterative Self-Organizing Data Analysis
LPDAAC	Land Processes Distributed Active Archive Centre
LULC	Land Use and Land Cover
MK	Man-Kendall
MODIS	Moderate-resolution Imaging Spectroradiometer
NASA	National Aeronautics and Space Administration
NGCC	National Geomatics Center for China
NGO	Non-Governmental Organization
NSE	Nash-Schiff efficiency
RMSE	Root Mean Square Errors
SHE	System Hydrologique European
SWAT	Soil and Water Assessment Tool
SWC	Soil and Water Conservation
TRMM	Tropical Rainfall Measuring Mission
USGS	United States Geological Survey
UTM	Universal Transverse Mercator
WetSpa	Water & Energy Transfer between Soil, Plants and Atmosphere
GPCP	Global Precipitation Climatology Project
RFEv2	African Rainfall Estimation Algorithm version 2
PERSIANN	Precipitation Estimation from Remotely Sensed Information using Artificial Neural Networks
GPCC	Global Precipitation Climatology Centre
GTS	Global Telecommunications System
GE	Geostationary satellites

LOE	Low Earth Orbiting satellites
AMSU	Advanced Microwave Sounding Unit
GOES	Geostationary Operational Environmental Satellite
TFWP	Trend-Free Pre-Whitening
MERET	Managing Environmental Resources to Enable Transition to more sustainable livelihoods
PSNP	Productive Safety Net Program
SLMP	Sustainable Land Management Project
EMA	Ethiopian Mapping Agency
SRTM	Shuttle Radar Topographic Mission
BoARD	Bureau of Agriculture and Rural Development
ATCOR-4	Atmospheric Correction for Airborne Imagery
ISRIC	International Soil Reference and Information Centre
UNCCD	United Nations Convention to Combat Desertification
FEWS NET	Famine Early Warning System Network
SEBAL	Surface Energy Balance Algorithm for Land
BFI	base flow index

Chapter 1

INTRODUCTION

1.1 BACKGROUND

The Tekeze Atbara river basin is one of the seven major international rivers originating in Ethiopia and flowing out to neighbouring countries and one of the largest tributaries of the Nile basin (Awulachew et al., 2007). The Upper Tekeze basin, located entirely in Ethiopia, forms the headwaters of this tributary and is currently the focus of the government of Ethiopia for economic development (Balthazar et al., 2013; Awulachew, 2010). To tackle the problems of recurrent drought and food insecurity, efforts have been made to harvest runoff water for irrigation and hydropower developments (Kifle and Gebretsadkan, 2017; Abraha, 2014; Haregeweyn et al., 2006; Yazew, 2005). The Ethiopian government has committed itself to an ambitious plan to eradicate food deficiency of the country in the shortest time possible by utilizing the available water resources (Awulachew et al., 2010; Yazew, 2005). However, land degradation, water scarcity and inefficient utilization of the available water resources are the key constraints for achieving this ambitious goal (Tadesse et al., 2011; Steenhuis et al., 2009; Kebede et al., 2006). Existing studies at regional and catchment level have shown that the availability of the water resources for economic development, especially in the semi-arid areas, is vulnerable to various natural and anthropogenic factors (Ayenew, 2007; Conway, 2005; Hurni et al., 2005).

Land use and climate change are the most dynamic factors that govern the variability of streamflow (Nepal et al., 2014; Tesemma et al., 2010; Li et al., 2009; Hurni et al., 2005). Alteration of existing management practices in a catchment influences the hydrological processes, including infiltration, groundwater recharge, base flow and surface runoff (Hurkmans et al., 2009; Li et al., 2009). The impact of human-induced changes on the hydrological processes differs from place to place, thus needing specific considerations in every circumstance (Haregeweyn et al., 2014; Lu et al., 2015). Climate variability/change is another factor that can significantly alter the timing, quantity and distribution of water in a basin (Gebremedhin et al., 2018; Mesfin et al., 2018; Gebrehiwot et al., 2011; Kim et al., 2008). Climate variability and climate change, as reflected in precipitation patterns, directly influence the availability of water resources in a basin (Zenebe, 2009; Seleshi & Zanke, 2004). The interaction between climate and a human-modified environment significantly affects the availability and distribution of water resources for economic development in a basin (Awulachew et al., 2010).

However, there is no clear understanding in the literature which factor is dominant in a given basin. For example, it has been reported in the scientific literature that the variation of hydrological flows in the Blue Nile basin is due to the change in patterns of rainfall over the basin (Kim et al., 2008; Kebede et al., 2006; Conway, 2005). In contrast, more recently, Tekleab et al. (2013), Gebremicael et al. (2013), Amsalu et al. (2007), Bewket & Sterk (2005), among others, found that human-made changes influenced the streamflow variations more significantly than climate variability. The effect of the human-modified environment on the water resources variability is reported to be even more pronounced in the Upper Tekeze basin (Abraha, 2014; Gebreyohannes et al., 2013; Gebrehiwot et al., 2011; Haregeweyn et al., 2006). Recent studies (e.g., Abraha, 2014; Bizuneh, 2013; Alemayehu et al., 2009; Zenebe, 2009) have indicated that the spatial variability of water resources in the Geba catchment, one of the headwaters of Upper Tekeze river basin has increased due to land use change coupled with limited and erratic distribution of rainfall.

The Ethiopian government has given strong attention to rehabilitate the degraded lands of the basin by introducing catchment management interventions (Gebremeskel et al., 2018; Nyssen et al., 2010; Alemayehu et al., 2009; Hengsdijk et al., 2005; Nyssen et al., 2000). As a result, a recognized success has been achieved in improving water availability at local level (Negusse et al., 2013; Schmidt & Zemadim, 2013; Nyssen et al., 2010; Alemayehu et al., 2009). The runoff water is trapped and infiltrates, which could be a potential source for increasing the availability of water. For example, Nyssen et al. (2010) found a reduction of surface runoff volume by 81% after management interventions in the My Zeg Zag watershed of the Geba catchment. Negusse et al. (2013) showed that groundwater availability in Arbiha Weatsbiha, a small watershed in the basin, increased more than ten times in the last 20 years. These changes are also reflected by the expansion of small-scale irrigated agriculture in the basin (Gebremeskel et al., 2018; Nyssen et al., 2010).

In contrast to the above success stories, other authors reported that catchment management interventions might have a significant negative role on the regional hydrological cycle (Wang et al., 2013; Mu et al., 2007; Xiubin et al., 2003). According to these reports, a reduction of annual flows and consequently water scarcity problems downstream are among the major potential negative influences of catchment management interventions. Although interventions can improve green water use efficiency and groundwater recharge at local level, the total surface runoff may reduce at a larger scale (Garg et al., 2012; Xiubin et al., 2003). Thus, understanding the precise impact of catchment management intervention and overall land use change on the downstream flow is critical for policymakers and catchment managers.

1.2   HYDROLOGICAL PROCESSES IN SEMI-ARID CATCHMENTS

Understanding the underlying hydrological processes is fundamental to develop a realistic modelling approach in simulating the actual physical characteristics of catchments. The hydrological process varies between semi-arid and humid environments (Pilgrim et al., 1988).

Because of climate variability in those areas, the availability of water resources is also known to be fluctuating at different spatio-temporal scales (Verma, 1979).

The major differences between hydrological processes of semi-arid and humid areas are well established in many studies (e.g., Hughes, 2008; Castillo et al., 2003; Yair & Kossovsky, 2002; de Wit, 2001; Bergkamp, 1998; Martinez-Mena et al., 1998). In general in arid and semi-arid areas, evapotranspiration is limited by the availability of water whereas the available energy is the main controlling factor in humid areas. Runoff generation in semi-arid areas is primarily controlled by surface properties rather than by the amount and intensity of rainfall (Yair and Kossovsky, 2002; Karnieli and Ben-Asher, 1993). The interaction among vegetation cover, microtopography and the hydrological response is perhaps more significant in semi-arid than in humid areas (Bergkamp, 1998; Pilgrim et al., 1988; Verma, 1979). Vegetation may increase the infiltration capacity of soils and reduce overland flows (Yair & Kossovsky, 2002; Pilgrim et al., 1988). Absence of vegetation cover in semi-arid catchments, in contrast, implies there is no protection of the soil to raindrop impact which can cause a reduction of infiltration capacity (Morin & Benyamini, 1977). In summary, less vegetation cover in semi-arid areas may lead to the absence of organic matter in the soil which can have significant effects on interception, infiltration, evapotranspiration and runoff response (Pilgrim et al., 1988).

Another factor that can significantly influence the hydrological processes of a catchment is the variability of precipitation. Precipitation in semi-arid areas tends to be more variable than in humid areas (Camacho et al., 2015; van de Giesen et al., 2005; Pilgrim et al., 1988). Sporadic high energy rainfall events generate almost all runoff (Love et al., 2010), and can cause soil erosion, reducing the infiltration capacity of the soil and thus enhancing surface runoff production in a catchment (Camacho et al., 2015; Wheater et al., 2007). Antecedent soil moisture content is recognized as one of the many runoff governing factors in semi-arid areas. However, the contribution of initial moisture in the soil to runoff generation is secondary as compared to surface properties and rainfall variability in the semi-arid catchments (Zhang et al., 2011).

In summary, infiltration excess runoff generation is common in degraded areas with low infiltration capacity, whereas the saturation excess process is dominant in less degraded areas with high infiltration capacity of the soil and improved land cover (White et al., 2011; Steenhuis et al., 2009; Cammeraat, 2004; Yair & Kossovsky, 2002; Martinez-Mena et al., 1998; Pilgrim et al., 1988). Infiltration excess overland flow is the most common runoff generation process in the semi-arid and arid areas where the rainfall intensity exceeds the infiltration capacity of the soil before it becomes saturated. In contrast, saturation excess runoff is the most dominant process in humid environments where surface runoff occurs only after the soil becomes saturated. It can be concluded that the assumption of linear hydrological response in wet areas does not hold in the semi-arid areas as the response of runoff in those environments is nonlinear (Pilgrim et al., 1998). Hence, recognizing those characteristics and differences is fundamental to develop a realistic modelling approach that can consider all unique characteristics of the semi-arid catchments.

1.3   IMPACT OF CATCHMENT MANAGEMENT DYNAMICS ON HYDROLOGICAL PROCESSES

The main challenges for ensuring food security in semi-arid areas are moisture stress, soil erosion, soil fertility decline and a shortage of pastures (e.g., Muys et al., 2014; Wang et al., 2013; Gebresamuel et al., 2010; Hengsdijk et al., 2005; Nyssen et al., 2004). To overcome these challenges, strong efforts have been made to rehabilitate the degraded lands of the Upper Tekeze basin through integrated watershed management interventions (Smit et al., 2017; Demissie et al., 2015; Nyssen et al., 2015a; Zhang et al., 2015; Zhao et al., 2013; Girmay et al., 2009). Many studies (Hurni et al., 2015; Nyssen et al.,2015a; Worku et al., 2015; Negusse et al., 2013; Lacombe et al., 2008) provide evidence that food production, land cover, soil fertility, surface and groundwater availability significantly improved after these interventions and erosion was reduced (Frankl et al., 2012; Tesfaye et al., 2012; Gebresamuel et al., 2010; Girmay et al., 2009). Nyssen et al. (2015b) showed that the recently observed small-scale irrigation intensification and increasing agricultural production and productivity in Northern Ethiopia was made possible because of catchment management interventions.

Soil and water conservation interventions implemented to protect land degradation can modify the hydrological processes of a catchment by changing the partitioning of the incoming rainfall at the land surface (Abouabdillah et al., 2014; Schmidt & Zemadim, 2013; Gates et al., 2011; Lacombe et al., 2008; Mu et al., 2007;). This can improve the availability of water during the dry season while decreasing the peak flow during the rainy season (Nyssen et al., 2010; Bewket & Sterk, 2005). Overall, large-scale catchment management interventions are capable of affecting the water resources of a basin both in positive and negative ways (Taye et al., 2015; Gates et al., 2011). Numerous studies (e.g., Abouabdillah et al., 2014; Schmidt & Zemadim, 2013; Nyssen et al., 2010; Lacombe et al., 2008; Mu et al., 2007) demonstrated that introducing of structural SWC measures such as terraces, stone bunds, soil bunds, trenches, check dams, percolation pits can reduce surface runoff and increase base flow. Similarly, biophysical SWC interventions, such as reforestation, residual moisture management, grass strips, conservation agriculture and enclosures, can also reduce surface runoff (Wang et al., 2013; Hengsdijk et al., 2005) and increase groundwater recharge.

Although SWC interventions have shown promising results with regard to improving the availability of water resource at the watershed level, the precise impact of large-scale implementation of these interventions on water resource availability for downstream users is less well known (Gebremeskel et al., 2018; Wang et al., 2013; Gates et al., 2011; Lacombe et al., 2008). The observed effect of SWC at the local level may not necessarily have the same proportional impact at larger scales. SWC interventions can improve green water use efficiency and groundwater recharge at a local level while total water outflows from treated catchments can reduce when compared to untreated catchments, which implies potential negative effects on downstream users (Gebremeskel et al., 2018; Garg et al., 2012; Gates et al., 2011). In contrast, numerous investigations (e.g. Negusse et al., 2013; Wang et al., 2013; Nyssen et al., 2010) have reported that subsurface flow from treated watersheds can contribute to increasing the total flow at the larger scale. Reconciliation of these contradicting findings requires a detailed study at various spatial scales.

1.4   HYDROLOGICAL MODELLING

Hydrological modelling is the application of mathematical expressions, which describe the quantitative relationship between input (e.g. rainfall) and output (e.g. runoff) (Arnold et al., 1998). The impact of past catchment management strategies and other human-modified environments can be identified using hydrological models (Abouabdillah et al., 2014; Jajarmizadeh et al., 2012; Wagener, 2007; Refsgaard, 1996). They can simulate the future potential impacts of land use management and climate change (Lu et al., 2015; Abraha, 2014; Goitom, 2012). Hydrological models can also provide a framework to conceptualize and grasp the relationship between climate, human intervention and water resources of a catchment for decision making and policy formulation (Beven, 2011; Legesse et al., 2003).

However, most hydrological processes are complex and almost impossible to analyse and study in terms of direct physical laws and conservation of mass (Jajarmizadeh et al., 2012; Beven, 2011). Difficulties in the correct representation of the different sources of variations make it unmanageable to apply the physical laws without simplifications (Githui, 2008). Considering these difficulties, hydrological models which represent the physical world in a simplified manner have been developed (Refsgaard & Knudsen, 1996). The accuracy of hydrological predictions in a watershed will always be limited by the simplified representations of the existing land characteristics in the models.

Top-down and bottom-up modelling approaches are the two basic modelling approaches that have been developed throughout the world for the understanding of hydrological processes and streamflow in a catchment (Savenije, 2009; Xu & Yang, 2010; Sivapalan et al., 2003). The prediction of catchment hydrological response in the bottom-up modelling approach is based on knowledge gained from the existing catchment physical processes (e.g. topography, climate, vegetation cover, soil) at relatively fine spatial and temporal scales, which is extrapolated to sub-basin and basin levels (Beven, 2011; Sivapalan et al., 2003). In contrast, the top-down modelling approach tries to understand the general characteristics of watersheds starting from the observed data using simple empirical models (Tekleab et al., 2011; Zhang et al., 2008; Sivapalan et al., 2003).

The bottom-up modelling approach requires a large amount of input data to represent each variable (e.g. topography, vegetation, climate, soil and other surface conditions) of the catchment. However, fully understanding the watershed hydrological processes is not achievable because not all input data are easily available (Suliman et al., 2015; Xu & Yang, 2010; Sivapalan et al., 2003). Calibration of parameters in the physically based models may lead to serious problems, such as scale issues, equifinality, non-uniqueness, and uncertainties about the calibrated model structure and the reliability of input data. Those models can suffer from over- parameterization and high prediction uncertainties that may also increase the tendency of divergence from the real world (Savenije, 2009; Sivapalan et al., 2003; Beven, 2002; Uhlenbrook et al., 1999). Physically-based distributed hydrological models (e.g. SWAT (Arnold et al., 1998), SHE (Abbott et al., 1986), WetSpa (Batelaan & De Smedt, 2001)) are the most popular bottom-up modelling approaches. Unlike bottom-up approaches, the top-down approaches use parametrically parsimonious models (Beven, 2011; Tekleab et al., 2011; Savenije, 2010). Representation of hydrological processes in a catchment is defined in a simplified way according to the perception of the user (Ampadu et al., 2013; Savenije, 2010). Such models have fewer parameters that represent only the dominant hydrological processes. Thus, model over-parameterization is not a primary concern as compared to physically based models because of the fewer number of degrees of freedom and their underlying interaction (Das et al., 2008). However, the top-down modelling approach requires the ability to define the net effect of small-scale interactions and feedback mechanisms to identify the hydrological processes at large scale and gradually reducing to processes at smaller scales (Sivapalan et al., 2003).

In conclusion, according to numerous comparative studies of model structures (e.g. Ampadu *et al.*, 2013; Beven, 2011; Daniel *et al.*, 2011; Das *et al.*, 2008; Reed *et al.*, 2004; Suliman *et al.*, 2015; Xu & Yang, 2010; Yang *et al.*, 2000), the semi-distributed and semi-lumped models outperform both fully distributed and fully lumped model structures. High model resolution and complexity does not improve hydrological simulations of a catchment. Similarly, simple lumped conceptual models are difficult to generalize since the most dominant processes in one location or scale may not necessarily be essential in another location or scale. For a better understanding of catchment processes, a change in hydrological investigation approach is needed (Beven, 2011; Daniel et al., 2011; Savenije, 2010; Das et al., 2008; Sivapalan et al., 2003). A combination of both approaches based on the existing facts and knowledge of the catchment creates a new hydrological modelling approach that can represent physical characteristics of a catchment that is parametrically efficient and applicable in hydrological data limited environments. The advantage of such an approach over simple lumped conceptual or fully distributed modelling approaches is, that it preserves the maximum simplicity of model structure while it considers the existing land characteristics (Beven, 2011; Savenije, 2010).

Nowadays, the development of a dynamic distributed hydrological model which requires little calibration to avoid over-parameterization and maximize available spatial data is becoming popular. The PCRaster/Python programming language frameworks (Karssenberg, 2010) are becoming important to develop dynamic and flexible distributed hydrological models such as Wflow (Schellekens, 2014), TOPMODEL (Beven, 2011; Gumindoga *et al.,* 2011) and FLEX-TOPO (Savenije, 2010). These hydrological models combine both physically and conceptually based modelling approaches to simulate hydrological responses from various contribution areas within the catchment. The dominant factor that determines the runoff generation in a catchment is topography which is closely linked to geology, soil, land use, climate, and ecosystem and, as a result, the dominant hydrological processes of a catchment. Such spatially dynamic hydrological models have the potential of simulating the impact of human-induced environmental changes (Hassabalah et al., 2017; Wang et al., 2016; Beven et al., 2011; Savenije, 2010; Sivapalan et al., 2003).

1.5 PROBLEM STATEMENT

The influence of human-made changes on the environment impacts the hydrological processes of the Upper Tekeze basin in Ethiopia. The basin is characterized not only by severe land degradation and moisture stress but is also known for its recent extensive integrated catchment management experience. It is claimed that catchment interventions have caused enhanced water availability at different locations in the upper parts of the basin (Gebremeskel et al., 2018; Nyssen et al., 2010). These interventions can also cause significant changes in the rainfall-runoff relationships. However, it is not accurately known how much the changes are and their downstream implications.

Few studies (Negusse et al., 2013; Nyssen et al., 2010; Alemayehu et al., 2009) have shown that the availability of surface and groundwater resources in the basin has significantly increased after the implementation of integrated catchment management. These achievements can also be evidenced by the expansion of small-scale irrigation schemes (Gebremeskel et al., 2018; Nyssen et al., 2010). However, none of these assessments has attempted to quantify the impact of catchment management measures on the hydrological dynamics at different scales and its consequences for downstream users. These few studies were done either at experimental plot level or in very small watersheds, from which it is difficult to extrapolate and infer basin-wide implications (Lacombe et al., 2008). Thus, an improved understanding of the effect of catchment management dynamics on the spatio-temporal variability of the hydrological processes and downstream flows is needed.

Understanding hydrological processes in changing environments in the semi-arid catchments of the Tekeze basin may inform catchment management interventions at different scales, both upstream and downstream of the basin. It is believed that a detailed analysis of the unique experience of the Upper Tekeze basin may be relevant for other semi-arid catchments elsewhere as well.

1.6 RESEARCH OBJECTIVES

The main objective of this study is to achieve a better understanding of the impact of catchment management dynamics on the overall hydrological processes and the spatial and temporal variability of streamflow in the Upper Tekeze-sub basin. The specific objectives are:

1. To evaluate and identify satellite-based rainfall estimates suitable for the Upper-Tekeze sub-basin;

2. To understand the linkages between rainfall and streamflow trends and identify possible drivers of change of the flow regime in the sub-basin;

3. To investigate the spatio-temporal dynamics of Land use/Land cover (LULC) and the associated land management changes;

4. To analyse the hydrological responses attributed to different land use types and their long-term dynamics in the sub-basin;

5. Modelling the low-flow modifications due to integrated catchment management interventions in the sub-basin.

1.7 STRUCTURE OF THE THESIS

The thesis is organized in eight chapters. The first chapter, provides a general introduction of this thesis including its justification and objectives

Chapter 2 describes the study area, including, topography, climate, hydrology, land use, soil, geological information and land management interventions in the basin.

Chapter 3 provides a validation of eight widely used satellite rainfall products at different spatio-temporal scales. The performance of these estimates was evaluated against 34 ground observations over the complex topography of the Upper Tekeze basin. The best performing product was used in the subsequent chapters.

Chapter 4 analyses rainfall and streamflow trends and identifies the drivers of streamflow changes in the study basin. Trend and change point detection of rainfall and streamflow were analysed using Mann-Kendall and Pettitt tests, respectively, using data records for 21 rainfall and 9 streamflow stations. The nature of changes and linkages between rainfall and streamflow were examined at monthly, seasonal and annual time scales.

Chapter 5 quantifies the human-induced landscape transformations in selected catchment of the study basin for the period of 1972-2014. A detailed land use/cover classification, accuracy assessment and change detection analysis was conducted and the main causes of the changes are identified and described in this chapter.

Chapter 6 analyses the hydrological response of dynamic land management changes in selected catchments of the study basin. The analysis was done using an integrated approach of a spatially distributed hydrologic model, indicators of hydrological alteration (IHA) and Partial Least Square Regression. A distributed hydrological model based on the Wflow-PCRaster/Python modelling framework was developed to simulate the hydrological response of land use/cover maps developed in chapter 5.

Chapter 7 analyses the impacts of soil and water conservation (SWC) interventions on the low-flow in two selected sub-catchments of the basin. The response of low-flows to the interventions were studied by comparing two catchments but also by comparing pre- and post-treatment interventions in both catchments. Calibrated model parameters were evaluated to verify to what extent the differences in catchment management interventions were reflected in the low-flow response, while the change of low-flow in both sub-catchments were assessed using Indicators of Hydrological Alteration (IHA).

Chapter 8 presents the research conclusions, its contribution to the scientific community, its limitations and recommendations for further investigations.

Chapter 2

STUDY AREA DESCRIPTION

2.1 LOCATION AND TOPOGRAPHY

The study area is located in the upper part of Tekeze-Atbara river basin in Northern Ethiopia between longitude 37.5° – 39.8° E and latitude 11.5° – 14.3° N (Figure 2.1). The Tekeze River originates in the southern part of the basin near the RasDeshen Mountains and flows in northern direction and then turns towards the west flowing into north-eastern Sudan, where the river joins the Atbara River (Zenebe, 2009; Belete, 2007). This study focuses on the Upper Tekeze basin which drains an area of 45,694 km^2 at the Embamadre gauging station (Figure 2.1). The mean annual flow at this point is 6.9×10^9 m^3/year, which is about 66 % of the total annual flow where the Atbara joins the main Nile. The basin is characterised by rugged topography consisting of mountains, highlands and terrains of gentle slopes. The elevation of the basin varies from 834 m.a.s.l. at the basin outlet in Embamadre to more than 4,528 m.a.s.l. in the Ras Dashen Mountains. Nearly half (46%) of the area is located between 1,000 to 2,000 m.a.s.l., and the remaining 30% and 24% of the area are located at an elevation of between 2,000 to 3,000 and above 3,000 m.a.s.l., respectively. This may indicate that topography could be a key factor in influencing microclimates in the basin.

2.2 CLIMATE

The basin is characterized by a semi-arid climate in the east and north and partly semi-humid in the south (Belete 2007). More than 85 % of the total annual rainfall falls in the wet season (June - September) which varies from 400 mm/year in the east to more than 1,200 mm/year in the south (Figure 2.1). The climate of the basin is dominantly semi-arid, with distinctive dry and wet seasons (Walraevens et al., 2015; Belete et al., 2007). The dry period over the region extends up to 10 months, and the maximum effective rainy season extends from 50 to 60 days (Gebremeskel and Kebede, 2018; Zenebe et al., 2013). The variations are mainly associated with the seasonal migration of the Inter-tropical convergence zone (ITCZ). The beginning and end of the ITCZ over the highlands of Ethiopia varies annually, which mostly causes the inter-annual rainfall variability (Nyssen et al., 2005; Seleshi and Zanke 2004). Chapter 3 will show that there is high spatial variability of rainfall within the relatively small Upper Tekeze basin.

The general pattern of rainfall over the basin is controlled by the complex topography which implies that the movement of air moisture can be substantially modified to create contrasting rainfall regimes in the region (Viste and Sorteberg 2013; Dinku et al., 2007; Huber et al., 2006). The sudden changes in elevation can obstruct the air mass movement to create a microclimate at the foothills or cause updraft over the mountains to create orographic rainfall (Dinku et al., 2007). Although in most regions rainfall increases with elevation due to the orographic uplifts (Moreno et al., 2014; Worqlul et al., 2014), this is not the case in the headwaters of the Upper Tekeze basin. Kiros et al. (2016) showed that rainfall over the Tigray region where the majority of the Upper Tekeze basin is located increases with elevation to the south while it decreases with elevation in the northern and north-eastern parts. Figure 2.2a shows the distribution of areal averaged rainfall over the basin. Compared with the topographic distribution (Figure 2.1), it is clear that rainfall increases with elevation in the south whilst it decreases with elevation in the northern and north-eastern parts of the basin. This implies that the relationship between rainfall and elevation is not uniform in the Basin (Figure 2.2b). This is attributed to the complex local topography, which alters proximity to the sources of moist air and seasonal movements of the ITCZ (Kiros et al., 2016; Van der Ent et al., 2010).

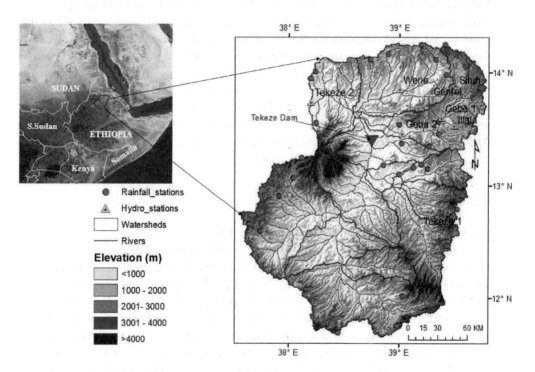

Figure 2.1. Location and distribution of rainfall and streamflow monitoring stations in the Upper Tekeze basin

During the rainy season, the ITCZ moves towards the Northern part of the basin, which brings moisture from the Atlantic and Indian oceans through westerly winds (Degefu et al., 2017; Mohamed et al., 2005). Westerly anomalies in the low-level circulation above Central Africa increase moisture transport from the Gulf of Guinea and the Indian Ocean (Viste and Sorteberge,

2013). When the moist air from these locations reach Central Africa, the westerly winds transport it to the Ethiopian highlands during the rainy season (Degefu et al., 2017, Viste and Sorteberge, 2013).When the rain-bearing winds reach the basin, their direction is modified by the local topography forcing the release of moisture in the lower areas before they reach the top of the mountains. This creates more intense and shorter duration convective rainfall events in the lowlands where warm and moist airflows encounter the mountain foothill which result in a low amount of rainfall in the highlands. Another possible reason for the low rainfall over the northern-eastern highlands is that whereas here the eastern rain-bearing winds are stronger, they carry less water vapour (Viste, and Sorteberg 2013). Van der Ent et al. (2010) showed that topography can play an important role in moisture cycling either by blocking or capturing moving air masses. This complex topography can also strongly influence the performance of satellite rainfall estimates (Haile et al., 2013). Algorithms used for rainfall estimations are challenged by a very cold surface and warm orographic rain over mountainous areas (Haile et al., 2013; Dinku et al., 2007)

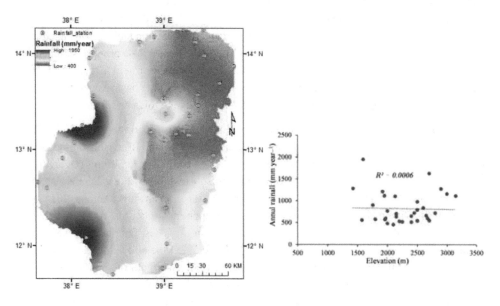

Figure 2.2. Distribution of annual average rainfall over the Upper Tekeze basin for the period of 1981-2015, (a) distribution of areal annual average rainfall (mm/year), (b) annual average point (station) rainfall against elevation of each station.

2.3 LAND USE

The dominant land use in the basin includes cultivable land (>70 %), open grassland, sparsely grown woodland, bushes and shrubs and exposed rocks (Zenebe, 2009; Tefera, 2003). The basin is characterized by severe land degradation through deforestation, overgrazing and cultivation on the rugged topography. Land use in most part of the basin is dominated by rainfed agriculture, main cops include Teff, wheat, barley, maize, sorghum and pulses, followed by shrubs, bare land, grassland and residential areas. However, irrigated agriculture at the household level and small-scale irrigation schemes have also increased significantly in the last 10 years in the eastern, northern and central parts of the basin (Nyssen et al., 2010; Alemayouh et al., 2009). Bare land and shrubs are the most dominant in the semi-arid eastern lowlands of the basin while most of the cultivable lands and very small forest coverage occur in the dry highlands of the basin. Agricultural and bare lands have expanded at the expense of all other land uses (Chapter 5).

2.4 SOIL AND GEOLOGICAL INFORMATION

The geology of the basin is dominated by limestone (27.6%) and metamorphic (27.2%), rocks, followed by sandstone (16%), limestone-marl (14.1%), dolerite (5.6%), shale (5.3%) and basalt (3.3%) (Zenebe, 2009). The drainage patterns are highly influenced by the foliation direction of Precambrian rocks in the northern and eastern part and neo-tectonic faults of Mekelle outlier in the central part of the basin (Abraha, 2014). The three major rock groups (sedimentary, igneous and metamorphic) are found well exposed in the eastern, northern and central parts of the basin (Birhane et al., 2016). Generally, the drainage system of the basin can be described as dendritic with some significant influence of major structures like folds and faults.

Major soil types identified in the basin includes, Eutric Vertisols on the level lands; Eutric Leptosols, Eutric Vertisols, Eutric and Calcric Cambisols and Haplic Luvisols on the sloping lands; Eutric Leptosols on the steep lands and Leptosols on composite landforms (Gebremeskel et al., 2018; Zenebe, 2009). Texturally, the dominant soil textures of the catchment are 40% clay loam, 30% sandy clay loam, 10% loam soil and 1% sandy loam soils (Abraha, 2014). Soil textures in the catchment are deeply weathered in the uppermost plateaus, rocky and shallow soils in the vertical scarps, coarse and stony soils on the steep slopes, finer textured soils in the undulating pediments and most deep alluvial soils are found in the alluvial terraces and lower parts of the alluvial deposits (Gebreyohannes et al., 2013). The depth of soils in the basin is limited due to contagious hard rocks and cemented layers. These soils are not suitable for crop production, but farmers have nevertheless been using them for cultivation due to the scarcity of arable land.

2.5 SOIL AND WATER CONSERVATION INTERVENTIONS

This basin is characterized by severe land degradation through deforestation, overgrazing and cultivation on the rugged topography. However, it is also known for its more recent experiences with soil and water conservation (SWC) activities (Nyssen et al., 2010; Alemayouh et al., 2009). SWC intervention plays an inevitable role in restoring the degraded landscapes including surface and groundwater resources (Gebremeskel et al., 2018; Pimentel, 1993). It has the capacity to sustainably maintain environmental and ecological services if properly implemented (Nyssen et al., 2015a). Traditional stone bunds have been commonly practised on cultivable land for many years in the highlands of Tekeze basin (Gebremeskel et al., 2018; Wolka, 2014). Construction of physical SWC structures in many landscapes have been given priority since the early 1990s (Asfaha et al., 2014; Nyssen et al., 2009; Gebremichael et al., 2005).

Some of the common physical SWC practices that have been introduced in many parts of the basin are terraces, stone and soil bunds, trenches and percolation pits, micro-basins and semi-circle terraces (Worku et al., 2015; Asfaha et al., 2014; Nyssen et al., 2010; Gebreegziabher et al., 2009; Herweg & Ludi, 1999). Furthermore, construction of water and soil harvesting structures such as check dams, sand storage dams, micro-dams, river diversions, ponds and shallow hand-dug wells (Nyssen et al., 2015b; Worku et al., 2015; Zeleke et al., 2014) have been given more attention in the last two decades. Recently, efforts have been made to integrate both biological and physical SWC measures. Biophysical measures include area enclosure and ex-closures (Belay et al., 2014; Muys et al., 2014; Nyssen et al., 2010; Descheemaeker et al., 2006b), plantation of multipurpose grasses and shrubs (Zeleke et al., 2014), enrichment plantation in upper catchments (Nyssen et al., 2015a; Alemayehu et al., 2009), agroforestry (Zeleke et al., 2014; Girmay et al., 2009), gully re-vegetating (Nyssen et al., 2015a; 2009) and soil fertility management (Araya et al., 2011; Alemayehu et al., 2009).

2.6 WATER RESOURCES

Streamflow in the Upper Tekeze basin is highly seasonal following the seasonality of precipitation. Most of the tributaries in the basin generate high runoff during the rainy season (June to September) and significantly decreases their flows or dry out in the long dry season (October to May). Hydrological flow measurements are very limited. Although the recording of flow data over the basin started in the late 1960s, it was discontinued for most of the gauging stations during the civil war in the 1980s. To this effect, only a few out of the total 39 stations have an extended period of more than 20 years of data. Table 2.1 presents the general characteristics of hydrological monitoring stations that have relatively consistent records.

The spatial and temporal variability of water, especially in the semi-arid areas of the basin is large. Consequently, the region is known for recurrent droughts and food insufficiency (Gebreyohannes et al., 2013; Zenebe et al., 2009). Even though some attempts have been made to harvest surface runoff for irrigation development, agriculture in the region remains largely rainfall dependent, which is characterized by low crops yield due to variation of rainfall both in amount and distribution (Gebrehiwot et al., 2011; Haregeweyn et al., 2006; Nyssen et al., 2005).

Except for one large hydropower dam which was inaugurated in 2009, no large-scale water resources development projects have been implemented in the Upper Tekeze basin. More than 92 micro-dams have been constructed by different stakeholders including Sustainable Agriculture and Environmental Rehabilitation in Tigray, Relief Society of Tigray and the regional government since the mid-1990s (Gebremeskel et al., 2018; Berhane et al., 2016; Hagos et al., 2016; Haregeweyn et al., 2006). However, the sustainability of benefits from these water harvesting structures has been threatened by siltation, leakage, structural damages, insufficient inflow and spillway erosion (Gebremeskel et al., 2018; Berhane et al., 2016).

Table 2.1: General Information of hydrological flow monitoring stations in the Upper Tekeze basin

Station name	Lat.	Long.	Altitude (m.a.s.l.)	Catchment area (km^2)	Recording period	Missing data (%)
Siluh	13.85	39.51	2,230	967	1973-2016	5.6
Illala	13.53	39.50	2,004	341	1980-2016	3.8
Genfel	13.80	39.60	1,997	733	1992-2016	3.1
Werie	13.85	39.00	1,380	1,770	1967-2016	52.8
Agula	13.69	39.58	1,994	692	1992-2016	1.3
Geba 1	13.6	39.38	1,748	2,445	1967-2016	49.0
Geba 2	13.46	39.02	1,370	4,590	1994-2016	1.2
Tekeze 1	12.60	39.19	1,490	1,002	1994-2016	4.2
Tekeze 2	13.74	38.20	845	45,694	1967-2016	55.0

Chapter 3

EVALUATION OF SATELLITE PRODUCTS IN THE UPPER TEKEZE BASIN [1]

Satellite rainfall products are considered as important options for acquiring of rainfall estimates in the absence of an adequate rain gauge network. However, estimates from these products need to be validated as their accuracy can be significantly affected by geographical position, topography and climate of specific region. Eight satellite rainfall products including, ARC2, CHIRPs, GPCP, CMorph, CMap, PERSIANN, RFEv2 and TRMM (3B42v7) were evaluated against ground observations over the complex topography of the upper Tekeze basin. The accuracy of the datasets was evaluated at different temporal and spatial scales over the period 2002-2015. The results show that the rainfall data of CHIRPS outperformed all other products at all temporal and spatial scales. Estimates from RFEv2, 3B42v7 and PERSIANN products are also closest to the measurements at rain gauges for all spatiotemporal scales: daily, monthly and seasonal, and both at the point and spatial scales. The remaining products performed poorly with PBIAS showing errors of up to 200% and lower r (<0.5) at all scales. However, the performance of all products improved as the temporal scale increased to month and season at all spatial scales. Compared to low altitudes <2,000 meter above sea level (m.a.s.l.), the PBIAS at high altitude (>3,000 m.a.s.l.) increased by 35% whilst r dropped by 28%. CHIRPS and 3B42v7 products showed the best agreement in mountainous terrains. However, all datasets show no consistency of the error sign. CMorph and 3B42v7 consistently overestimate rainfall relative to all rain gauges using a pixel-to-point rainfall comparison approach and in lowland areas during the areal averaged rainfall comparison. The other six products showed a clear underestimation at all spatial scales. In summary, the results show that rainfall estimates by CHIRPS, RFEv2 and 3B42v7 have a consistently better agreement with ground rainfall than other products at all spatiotemporal scales. Considering the complex topography and limited gauges, the performance of CHIRPS, RFEv2 and 3B42v7 indicates that these products can be used for hydrological and overall water management applications in the region.

---

[1] Based on: Gebremicael, T.G., Mohamed, Y.A., van der Zaag, P., Gebremedhin, A., Gebremeskel, G., Yazew, E., Kifle, M., (2019b). Evaluation of multiple satellite rainfall products over the rugged topography of the Tekeze-Atbara basin in Ethiopia. *Int. J. Remote Sens.*, 1-20. DOI: https://doi.org/10.1080/01431161.2018.1562585

3.1 INTRODUCTION

Accurate information on rainfall data is necessary for water management, hydrological applications, and agricultural forecasts (Wang et al., 2017; Guo and Liu, 2016; Sunilkumar et al., 2015). Despite its importance for socio-economic development, ground-based rainfall measurements are sparse and unevenly distributed, especially in developing countries (Behrangi et al., 2015; Gebremichael et al., 2014; Haile et al., 2010). The recommended density of ground rainfall measuring network in tropical regions is one gauge per 600 - 900 km^2 for flat and 100 – 250 km^2 for mountainous areas, respectively (WMO, 1994). However, such recommendations are not available in most tropical regions (Worqlul et al., 2014; Taye and Willems, 2013). E.g., it is only one gauge per 1400 km^2 in our case study of the Upper Tekeze basin. Satellite rainfall products are increasingly becoming an important source of rainfall data, in particular in poorly or (un)gauged catchments (He et al., 2017; Katsanos et al., 2015; Meng et al., 2014).

Satellite rainfall products are available with almost global coverage and are becoming cost-effective for hydrological applications (Meng et al., 2014; Thiemig et al., 2012).The spatiotemporal resolutions and measurement accuracy of these products are continuously improving because of the advancement in sensor technologies and estimation techniques. A number of high resolution rainfall products are now available at quasi-global scale (Wang et al., 2017; Xu et al., 2017; Behrangi et al., 2015). The Tropical Rainfall Measurement Mission (TRMM), African Rainfall Estimation Algorithm version 2 (RFEv2), African Rainfall Climatology (ARC), Global Precipitation Climatology Project (GPCP), Precipitation Estimation from Remotely Sensed Information using Artificial Neural Networks (PERSIANN) Climate Hazards Group InfraRed Precipitation with Stations (CHIRPS) and CPC Morphic technique (CMorph) are among the common products that have been widely applied globally (Dembélé and Zwart, 2016; Guo and Liu, 2016; Haile et al., 2010; Dinku et al., 2007).

However, satellite rainfall products need to be validated as their accuracy can be affected by geographical position, topography, and climate as well as by the algorithms used to derive rainfall data from satellite measurements (Meng et al., 2014; Xue et al., 2013). Several studies on the validation and comparisons of these products with ground measurements have been conducted at different scales (e.g., Guo, and Liu 2016; Worqlul et al., 2014; Jiang et al., 2012; Dinku et al., 2007). However, their performance varies for the same data type across different regions and seasons (Toté et al., 2015; Gebremichael et al., 2014; Hu et al., 2014). This indicates that the performance of satellite products largely depends on the location, topography, season, and hydro-climatic characteristics of the basin. Therefore, the reliability of satellite rainfall needs to be validated and compared against ground measurements to a specific area and temporal scales before it can be used in any subsequent application (Xu et al., 2017; Ouma et al., 2012; Feidas, 2010).

A number of studies have been conducted in Ethiopia to evaluate satellite rainfall products (Bayissa et al., 2017; Haile et al., 2013; Gebremichael et al., 2014; Worqlul et al., 2014; Dinku et al., 2007). However, these studies have been mainly focused on the Upper Blue Nile basin and to some extent on central Ethiopia. In the Upper Tekeze basin, there have been no comprehensive validation studies. Therefore, this study aims to validate eight of the widely used satellite rainfall products at different spatiotemporal scales. The relationship between satellite rainfall products and topography is carefully explored based on categorical schemes in order to understand possible errors produced by the rugged terrain.

3.2 DATA AND METHODS

3.2.1 Rain gauge data

The ground rainfall data used for validation of the satellite products comprised of 34 stations located within and surrounding the basin (Figure 2.1). These data were provided by the Ethiopian Meteorological Service Agency (NMA). The dataset includes daily data for the period from 2002 to 2015. Although the number of stations is relatively good, their distribution over the basin is not uniform. Most of the gauges are located in easily accessible areas and the distribution of gauges in the lowland areas are sparse (Figure 2.1). Interestingly, most of the rainfall stations with a relatively good quality of data are located in the highland areas where the spatial variability of rainfall is high. A summary of these ground measurements with vertical locations is presented in Appendix A (Table A-1).

The rainfall data from each station was scrutinized on outliers and missing values. All outliers were then compared to neighbouring gauges to cross-check if observed extreme values resulted from extreme climate events. Stations with large data gaps were excluded from the analysis. After data screening, 34 out of the 75 stations in the basin were found to be reliable with a better record length of data. It is to be noted that the rain gauge stations are not part of the Global Precipitation Climatology Centre (GPCC) network used for calibration of satellite products.

3.2.2 Satellite rainfall products

Eight satellite rainfall products were used for this study (Table 3.1). These products were selected based on the following criteria: public domain dataset, long-term data availability, reasonable spatiotemporal resolution, near-real-time availability and their wide applicability in Africa (Dembélé and Zwart, 2016; Thiemig et al., 2012; Dinku et al., 2007).

Table 3.1: Summary of selected satellite rainfall products for this study (in descending order of spatial resolution)

Product	Temporal resolution	Spatial resolution	Coverage	Starting date
ARCv2	Daily	0.1°	40°N-40°S,20°W-55°E	1983
CHIRPSv8	Daily	0.05°	50°N-50°S,0°-360°E	1981
CMap	pentad	2.5°	Global	1998
GPCP	Daily	2.50°	Global	1979
CMorph	3 hourly	0.25°	Global	2002
RFEv2	Daily	0.1°	40°N-40°S,20°W-55°E	2001
PERSIANN	Daily	0.25°	Global	1983
TRMM3B42V7	3 hourly	0.25°	50°N-50°S,0°-360°E	1998

The CHIRPS datasets, developed by the US Geological Survey (USGS) and the Climate Hazards Group at the University of California are blended products which combine global climatologies, satellite observations, and in-situ rainfall observations from Global Telecommunications System (GTS) (Funk et al., 2014; Knapp, 2011). CHIRPS product is a third generation rainfall estimation procedure based on several interpolation techniques to create continuous grids from raw point data (Funk et al., 2014; Knapp, 2011). The input data for CHIRPS product development includes Climate Hazard Precipitation Climatology, Quasi-global geostationary Thermal infrared satellite observation from National Oceanographic and Atmospheric Administration (NOAA), National Climate Data Center, Atmospheric model rainfall from NOAA Climate Forecast Systems and TRMM 3B42 product from NASA (Mohamed et al., 2005). CHIRPS incorporates 0.05° resolution satellite rainfall estimates from the various sources with ground data to produce daily time series (Dembélé, and Zwart 2016). Such products have been applied in many regions, including East Africa (Shukla et al., 2014), the Nile basin (Knapp, 2011) and Cyprus (Dembélé and Zwart, 2016).

The RFEv2 is provided by the National Oceanographic and Atmospheric Administration Climate Prediction Center (NOAA-CPC) for Famine Early Warning Systems Network (FEWS-NET) to assist in disaster-monitoring activities over Africa (Herman et al., 1997). RFEv2 has been operational since 2001 and uses rainfall estimates from Passive microwave (PM) and infrared (IR) sensors which are the most widely used electromagnetic spectrum channels and data from meteosat and daily rainfall from the GTS reports (Dinku et al., 2007). RFEv2 is a merging technique of Geostationary satellites (GEO) and Low Earth Orbiting satellites (LEO) which increases the accuracy of rainfall estimates (Xie, and Arkin, 1997). Data from the Global Precipitation Climatology Center (GPCC) are used to reduce the bias of the product.

ARCv2 is also produced by the NOAA-CPC and provides daily rainfall data over Africa. The ARCv2 uses input from GEO IR data centred over Africa with quality control using GTS gauge observation from Africa (Dembélé and Zwart, 2016). The only difference between RFEv2 and

RCv2 is the use of polar-orbiting PM and Geostationary IR data. The ARCv2 does not include PM estimates and uses only 3 hourly IR instead of 30 minute data (Dembélé and Zwart, 2016).

CMorph is also a product from NOAA-CPC. Unlike the other products, the CMorph product is not an algorithm for merging of the PM and IR estimates but only uses the IR information for the spatial and temporal evolution of clouds, not the rainfall estimates (Asadullah et al., 2008; Joyce et al., 2004). It uses rainfall estimates derived from low orbit PM observations and propagates these features using high temporal and spatial resolution IR data (Joyce et al., 2004). According to Dinku et al. (2007), the CMorph combines the superior retrieval accuracy of the PM and higher resolution of IR data. This method is highly flexible as it allows incorporation of any rainfall estimate from PM satellites. A detailed description of this product is found in Haile et al. (2013) and Joyce et al. (2004).

The PERSIANN precipitation estimates are developed by the Center for Hydrometeorology and remote sensing at the University of California (Ashouri et al., 2015). The TRMM satellite, GEO and IR are input datasets of this product (Hsu et al., 1997). It uses an artificial neural network approach to merging the IR and PM data, and the rainfall estimates are based on the infrared brightness temperature image provided by GEO satellites (Hsu et al., 1997). The neural network is calibrated with Special Sensor Microwave Imager (SSM/I), TMI and Advanced Microwave Sounding Unit (AMSU) data which is then used to acquire rainfall estimates using IR data as a basis (Sapiano and Arki, 2009).

The TRMM product (3B42V7) was developed by the National Aeronautics and Space Administration (NASA). It is obtained from the TRMM Multi-satellite Precipitation Analysis (TMPA) algorithm which combines IR and PM data retrievals (Huffman et al., 2007). TRMM uses GEO and LEO satellites in which the rainfall is estimated from various PM sensors (Worqlul et al., 2014; Hessels, 2015). The TRMM satellite makes the data available both in the near-real-time and delayed research quality. It incorporates gauge data for bias correction from several sources including national and regional meteorological services (Funk et al., 2014). The 3B42v7 product of TRMM was aggregated from the TMPA-3B42 3-hourly estimates and merged with station data to produce daily rainfall estimates (Dinku et al., 2007; Huffman et al., 1997). The 3B42V7 estimates achieve a relatively higher accuracy among the different TRMM products (Haile et al., 2010). Rainfall estimates from TRMM showed a good agreement with the ground measurements over the Upper Blue Nile sub-basin (neighbouring to Tekeze basin) of the Nile Basin (Gebremichael et al., 2014; Haile et al., 2010).

The GPCP is a blended product which combines the Global Precipitation Climatology Center (GPCC) gauge data with the PM and IR rainfall estimates (Huffman et al., 1997). The PM estimates in this product are based on SSM/I data from the Defence Meteorological Satellite Program (DMSP, US) while the IR data come mainly from Geostationary Operational Environmental Satellite (GOES) Precipitation Index (PI) (Xie and Arkin., 1995). This technique is advantageous as it combines rainfall estimate information from many data sources by taking the strength of each data type. The microwave estimates in GPCP are used to adjust the IR estimates and to create multi-satellite rainfall data. The value at each grid box of the multiple satellite product is adjusted by a large-scale gauge average rainfall (Feidas, 2010). The product suffers from inhomogeneity in addition to the coarse spatial resolution (Dinku et al., 2007). Although relatively coarse in time and space, this product has proven to be useful for large-scale climate studies (Sapiano and Arki, 2009).

CMap products include monthly and pentad (5-day) mean rainfall estimates at 2.5° spatial resolution (Feidas, 2010). These techniques produce rainfall estimates by merging ground station data with rainfall estimates from several satellite-based algorithms (Xie and Arkin, 1997). As described in Xie and Arkin (1997), inputs are derived by combining of GEO and polar orbiting infrared, PM retrievals and rain gauge observations. First, the IR and PM rain estimates are merged using a maximum likelihood approach where the estimate with weights are derived by comparison to the gauge analysis. Then, the gauge analysis is used to obtain an absolute value of the merged product (Feidas, 2010). The combined satellite estimates are assumed to represent the structure of rainfall distribution with minimum bias in rain gauge estimates (Dinku et al., 2007). Similar to GPCP, the CMap product suffers from the coarse spatial resolution (Feidas, 2010)

3.2.3 Evaluation and Validation processes

The spatial patterns of eight satellite products were evaluated and compared against rain gauge data at daily, monthly, and seasonal scales. Mmajor seasons in the study area are the rainy season (June–September), dry season (October–February) and short rainy season (March–May) (NMSA, 1996). As more than 85% of the total annual rainfall occurs during the wet season (June-September) (Gebremicael et al., 2017), the analysis for the seasonal scale only considered this period. Considering the given climatic variability, complex topographical characteristics and hydrological working units of the basin, the performance of these products were evaluated using two approaches, namely point-to-pixel and areal averaged rainfall comparison.

Rainfall over a complex topography like the Tekeze basin is characterized by a high spatial variability within a small area, which implies that evaluation of such satellite products should be at the smallest possible spatial and temporal scales (Thiemig et al., 2012). Accordingly, in the first approach, all satellite rainfall products from the corresponding grid cell were compared to the ground observed data within the satellite grid. This comparison approach is essential to understand the performance of these products at a smaller scale and complex terrains with varying climate. However, it may not represent the conventional working unit for hydrological applications (Dembélé and Zwart, 2016; Katsanos et al., 2015). The variance of satellite estimate is smoother in space and time as these products are represented by the spatial averages over the pixels. For this analysis, the satellite rainfall products were extracted for the location of each rainfall station and their performance was evaluated using statistical indices. It was assumed that the amount of point rainfall is uniform in the area of the pixel which may not necessarily be true for coarser resolution satellite products. To account for the effect of topography, the performances of these products were also compared based on categorical validation schemes at different elevations (1,000 – 2,000, 2,000 – 2,500, 2,500 – 3,000 and 3,000 – 4,500 m.a.s.l.). The second approach was based on the areal rainfall comparison at different spatial scales. Spatially, satellite products were validated at sub-basin and basin level by comparing spatially aggregated pixel values against a corresponding spatially interpolated (1km x 1km resolution) observed rainfall from gauge stations. Representative sub-basins from lowland (with an average elevation of stations 1,400 m.a.s.l.) and highland (with an average elevation of 3,000 m.a.s.l.), were considered to account the topographic effects. Point gauge rainfall data were interpolated into areal rainfall using kriging interpolation method. The kriging interpolating technique was adopted because it produces reasonable spatial maps with better accuracy in rugged terrains (Goovaerts. 2000). However, it is also important to note that interpolating of sparse and unevenly distributed rain gauges over the complex terrains of the basin may also introduce uncertainties in the comparison processes.

3.2.4 Evaluation statistics

The satellite rainfall products were quantitatively evaluated against ground observations using four statistical indices: the relative percent of bias (PBIAS), Pearson correlation coefficient (r), Root Mean Square Error (RMSE), and Mean Absolute Error (MAE) (Table 3.2). The RMSE and MAE evaluate the average magnitude error between satellite estimates and gauge measurements. The MAE is suitable to describe uniformly distributed errors, while the RMSE is more appropriate if the errors are normally distributed (Chai and Draxeler, 2014; Derin and Yilmaz 2014). A detailed description of these indices can be found in Toté et al. (2015) and Thiemig et al. (2012). The lower the PBIAS, RMSE and MAE values, the closer the satellite estimates are to the ground measurements. The unit of RMSE and MAE is mm/time period.

Table 3.2: Statistical indices used for the satellite rainfall products performance evaluation

Statistical measure	Equation
Root Mean Square Error (RMSE)	$\sqrt{1/N(\sum_{i=1}^{n}(y_i - x_i)^2)}$
Mean Absolute Error (MAE)	$1/N(\sum |y_i - x_i|)$
Percentage of bias (PBias)	$((\sum y_i - \sum x_i)/\sum x_i) \times 100$
Pearson correlation coefficient (r)	$\dfrac{\sum(x_i - \bar{x})(y_i - \bar{y})}{\sqrt{(x_i - \bar{x})^2}\sqrt{(y_i - \bar{y})^2}}$

Where x_i is observed rainfall from rain gauge, y_i is satellite rainfall product, N is the number of pairs of products, $\bar{x}$ and $\bar{y}$ are the average of observed and satellite rainfall data, respectively.

3.3 RESULTS AND DISCUSSIONS

3.3.1 Comparison at pixel-to-point spatial scale

The performance of satellite estimates was evaluated by comparing these with data from 34 rainfall stations at the grid level covering the location of the station. Comparisons were carried out at daily, monthly, and seasonal periods. The box plots in Figure 3.1 shows the summary of statistical indices calculated by comparing the daily rainfall estimates. The box plots display the full range of variation from minimum to maximum (vertical line) and the quartiles and median (horizontal lines) for all satellite estimates. The PBIAS (Figure 3.1a) and correlation coefficient (Figure 3.1b) of each satellite estimates indicate that the daily estimates performed poorly in the majority of stations. However, CHIRPS, RFEv2 and 3B42V7 had a relatively better performance with lower PBIAS, RMSE and MAE and higher r compared to other products (Figure 3.1a-d). The values of each statistical index for all stations are summarized in Appendix A (Tables A-2 to A-5). The average value of PBIAS for all stations was -13% (ranging from -63% to +26%), -16% (-88% to +46%) and 17% (-39% to +86%) for CHIRPS, RFEv2 and 3B42V7, respectively. Similarly, r value of these products was ≥ 0.5 in the majority of stations with a range (average) value of 0.36 to 0.68 (0.50), 0.16 to 0.57 (0.51) and 0.25 to 0.62 (0.49), respectively. Lower values of RMSE and MAE were also observed for CHIRPS, RFEv2 and 3B42V7 products compared to the others (Tables A-4 and A-5).

Moreover, the standard deviation of these measuring matrices supports this result with lower values for the CHIRPS, RFEv2 and 3B42V7 products. The remaining products failed to capture the observed daily rainfall with $r < 0.5$ and higher PBIAS, RMSE and MAE in most stations. The effect of topography on the satellite rainfall products was also further validated based on categorical schemes. As shown in Table 3.3, the daily PBIAS calculated for the different elevations of all products continuously increased as we go up towards a higher elevation. TRMM and CMorph show a consistent overestimation of rainfall across all elevation categories whilst all remaining products except GPCP and RFEv2 underestimated in all elevation ranges. RFEv2 and GPCP also underestimate in all elevation ranges except in 2,000-2,500 and 2,500-3,000 m.a.s.l, respectively, which overestimated the rainfall.

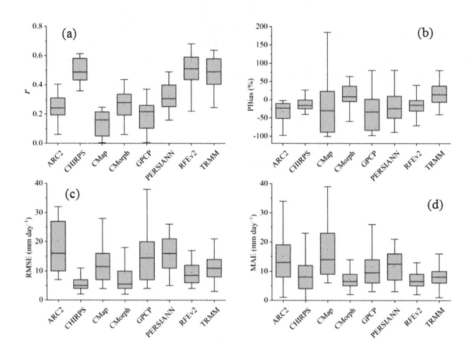

Figure 3.1: Comparison of daily satellite rainfall estimates with ground measurements, (a) Correlation (%), (b) PBIAS (r), (c) RMSE and (d) MAE

Table 3.3: PBias comparison between various satellite datasets and gauge observations for different elevation categories valid on the daily temporal scale

Elevation (m)	ARC2	CHIRPS	CMap	CMorph	GPCP	PERSIANN	RFEv2	TRMM
1,000–2,000	-26	-5	-23	8	-25	-7	-9	9
2,000–2,500	-31	-10	-34	20	29	-17	-24	12
2,500–3,000	-36	-14	-43	27	-45	-29	27	21
3,000–4,500	-48	-24	-75	33	-57	-35	-43	25
Average	-35	-13	-44	22	-24	-22	-12	17

The accuracy of these products to reproduce the observed rainfall was further investigated at monthly time series that is by aggregating data from daily to monthly time step. Table 3.4 shows improved values of accuracy indicators obtained for the monthly time step. The correlation for CHIRPS, RFEv2 and 3B42V7 were >0.5 in all stations with an average value of 0.61, 0.59 and 0.56, respectively. The RMSE and MAE indices also decreased at the monthly scale, which implies the agreement between satellite and ground rainfall increased (Table 3.4). CHIRPS, RFEv2 and 3B42 outperformed the other products. ARCv2, CMap and GPCP again performed poorly with r < 0.5 and higher PBIAS (Table 3.4). Moreover, comparisons based on average monthly point rainfall (2002-2015) at the given locations indicate that rainfall estimates of CHIRPS, RFEv2 and 3B42V7 products agree with the corresponding ground measurements (Figure 3.2). Monthly rainfall patterns from those products have a consistent and strong agreement with the ground rainfall compared to the remaining products.

Table 3.4: Range of various statistics from a comparison between various satellite datasets and gauge observations valid for the monthly temporal scale

Satellite product	RMSE (mm/month)	MAE (mm/month)	PBias (%)	r
ARCv2	35–222	18–122	-54--8	0.01–0.54
CHIRPS	4–49	20–43	-20–34	0.55–0.71
CMap	42–216	26–86	-48–385	0.21–0.52
CMorph	37–284	22–51	-76–50	0.11–0.68
GPCP	33–225	20–122	-50–27	0.21–0.57
PERSIANN	22–57	27–58	-30–53	0.42–0.61
RFEv2	20–56	18–50	-22–26	0.52–0.62
TRMM	24–133	12–54	-22–44	0.53–0.64

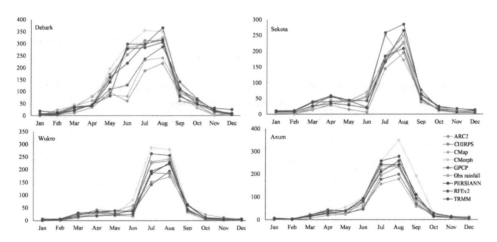

Figure 3.2: Comparison of observed mean monthly rainfall (2002-2015) with estimated satellite products at four representative ground stations

To gain further information on the skill of the satellite retrievals at the seasonal scale, comparisons were also made for the entire rainy season (June-September). Figure 3.3 presents an inter-comparison of wet season satellite rainfall estimates with the observed rainfall of the same period using the range value of correlation coefficient (r). The spatial distribution of correlation coefficients (Figure 3.3 and PBIAS (Figure A-1) indicated by the range values shows that six satellite products had an excellent agreement with ground rainfall (CHIRPS, 3B42V7, PERSIANN, RFEv2, ARC2, and CMorph). Compared to the daily (Figure 3.1) and monthly (Table 3.4) analysis, the performance of all products significantly improved when considering the entire wet season (Figure 3.3). CHIRPS, RFEv2, and 3B42V7 correlated best with the observed rainfall compared to the remaining products with r values ranging from 0.56 − 0.95, 0.38 − 0.94 and 0.43 − 0.89, respectively. The PBIAS, RMSE and MAE indices of these products are also lower than the other products (Tables A-6 and A-7). Although good correlation coefficient values were obtained in several stations (Figure 3.3), the average values of RMSE and MAE for CMorph were very high, suggesting that the errors across the different stations are not uniformly distributed (Table A-6 and Table A-7). The PERSIANN and ARCv2 products showed a reasonable agreement with gauged rainfall, while the GPCP and CMap showed poor agreement (Figure 3.3 and Tables A-6 and A-7).

Comparison at different temporal scales showed that all satellite products suffer from both over- and under-estimations. The negative and positive values of PBIAS shown at different temporal scales (e.g., Table 3.3, 3.4 and Figure 3.1b) explains under- and over-estimation, respectively. 3B42V7 and CMorph systematically overestimate the rainfall in more than 20 stations while they underestimate rainfall in the remaining stations. The RFEv2, GPCP, ARC2 and CMap products consistently underestimate rainfall in the majority of ground stations.

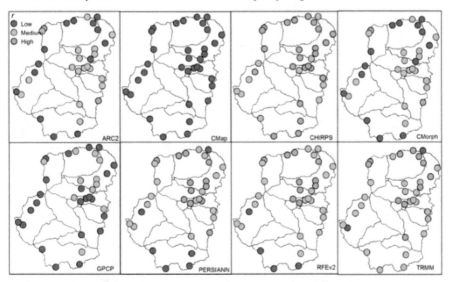

Figure 3.3: Spatial distribution of correlation coefficients (r) for the wet season comparison. Low (r = 0.0 − 0.50), Medium (r = 0.50 − 0.80) and High (r = >0.80).

It is important to note that the inconsistent estimation of rainfall by all products is likely due to the effect of the rugged terrain (see Table 3.3). The overall performance of the satellite rainfall products is lower in the peripheries of the basin where most stations are located in the mountainous area with an elevation of > 2,500 m.a.s.l. (Table 3.3). For example, Figure 3.4 compares the pattern of statistical indices for all products in four representative stations, highland (3,000 m.a.s.l.), lowland (1,400 m.a.s.l.), and midland (2,000 m.a.s.l.). A lower correlation and higher PBIAS are observed for the majority of the satellite estimates in Debark compared to Sekota with elevations of 3,000 and 1,960 m.a.s.l., respectively. A relatively better performance occurred in the central, eastern, and north-western parts of the basin where stations are located below 2,500 m.a.s.l. In addition, similar to the ground measurements, the relationship between satellite rainfall and elevation is not straightforward over the basin. This non-uniform pattern is due to the complex local topography which may affect the satellite's ability to detect light rainfall events (Bharti and Singh 2015). The correlation of these products showed a poor agreement at a higher elevation. A similar study in the neighbouring Upper Blue Nile basin by Gebremichael et al. (2014) also showed that satellite products failed to capture the ground rainfall in mountainous compared to lowland areas. This result is consistent with other studies carried out elsewhere (e.g., studies such as Kimani et al., 2017; Guo, and Liu 2016; Hu et al., 2014).

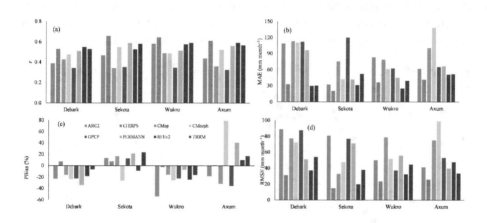

Figure 3.4: Monthly statistical indices at pixel to point rainfall comparison. Debark station is located in the highland, Sekota in the lowland while Wukro and Axum are located at intermediate heights.

In summary, CHIRPS, RFEv2 and 3B42V7 products demonstrate the best agreement with the observed rainfall across the basin compared to other products for the daily and monthly comparisons (Figure 3.1 and Table 3.4). The evaluation indices at both daily and monthly timescales show that CHIRPS performs best among the used products, followed by RFEv2 and 3B42V7 during daily and 3B42V7 and RFEv2 during the monthly comparison. Similarly, CHIRPS consistently outperformed the other products at seasonal (wet) comparison followed by PERSIANN, RFEv2 and 3B42V7 products (Figure 3.3). This implies the best performing products next to CHIRPS varied with timescale. The errors (PBIAS) and correlation with ground rainfall of such products were within ±25 and >0.5 in all spatiotemporal scales. All products except ARC2, GPCP and CMap showed a good agreement (PBIAS < ±10) at elevations <2,000m.a.s.l. with the observed rainfall. However, CHIRPS and 3B42V7 are the only products that exhibit a good performance for the high altitude areas lying between 3,000-4,500 m.a.s.l. The PBIAS of both products was within ±25 whilst the remaining products performed poorly as the elevation increases with PBIAS up to -75% in altitudes > 3,000 m.a.s.l. This implies that only these products could be useful for any hydrological applications in the basin. Over- and under-estimation of daily, monthly and seasonal rainfall by CHIRPS was smaller compared to the other products. The performance of all satellite products improved significantly when aggregated from daily to monthly and seasonal timescales.

3.3.2 Comparison based on areal araveged rainfall

Areal rainfall of each product was also compared at different spatiotemporal scales with the corresponding interpolated rainfall from 34 rain gauges. Table 3.5 shows the performance of all products at the basin level. The CHIRPS, RFEv2, 3B42V7, PERSIANN, and CMorph areal rainfall estimates had the best accuracy at all temporal scales. However, similar to point comparison, all products showed a poorer performance at daily compared to monthly and seasonal timescales (Table 3.5). Poorer performance at a daily scale can be explained by erroneous (non-detection) of more localized convective rainfall events. Higher accuracies at larger timescales are due to the fact that the errors at a smaller timescale are symmetrical and offset each other when aggregated. The CMap, ARCv2, and GPCP continued to show poor performance at all timescales. Except for 3B42V7 and CMorph, all products underestimated rainfall at all temporal scales (Table 3.5). The 3B42V7 and CMorph products consistently overestimated rainfall.

Table 3.5: Comparison of satellite and observed areal rainfall, at basin scale, at different time scales

Index	Temporal scale	ARC2	CHIRPS	CMap	CMorph	GPCP	PERSIANN	RFEv2	TRMM
PBias	Daily	-58	-10	-38	18	-41	-13	-8	11
	Monthly	-41	-8	-33	15	-28	-11	-6	8
	Wet season	-21	-6	-24	11	-19	-8	-3	6
r	Daily	0.22	0.49	0.21	0.19	0.15	0.28	0.48	0.41
	Monthly	0.39	0.69	0.28	0.41	0.3	0.54	0.5	0.56
	Wet season	0.41	0.88	0.39	0.55	0.36	0.64	0.72	0.7
RMSE (mm)	Daily	111	36	60	42	89	48	50	25
	Monthly	178	42	80	70	205	49	61	34
	Wet season	201	79	309	112	395	137	67	142
MAE (mm)	Daily	102	19	54	39	79	48	25	25
	Monthly	141	41	78	68	131	51	36	31
	Wet season	198	71	229	102	131	171	68	123

Whereas all satellite rainfall estimates showed a consistent improving pattern with a temporal scale, their performance did not show a uniform pattern with increasing spatial scale. Figure 3.5 compares the average correlation of the products at different spatial scales for the wet season. Most products performed worse at the basin scale compared to the pixel-to-point and lowland sub-basin scales. The likely reason could be that the areal averaged rainfall over the complex topography may suffer from limitations due to the uneven distribution of rain gauges. Unlike the others, the performance of CHIRPS, 3B42V7, and CMap improved at basin level compared to pixel-to-point scale (Figure 3.5). The relatively poor performance at the basin scale for most products is likely due to the topographical variations across the basin. Variations of topography can also significantly compromise the interpolation of observed rainfall (Thiemig et al., 2012). The rainfall stations are also sparse and unevenly distributed over the basin, which can be a source of systematic errors when interpolating areal rainfall (Dembélé and Zwart, 2016; Toté et al., 2012).

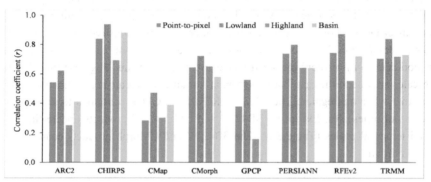

Figure 3.5: *Comparison of seasonal averaged correlation at a pixel, sub-basin, and basin scales*

To further understand the effect of complex terrains on the performance of the satellite products, the seasonal areal rainfall of representative highland and lowland sub-basins was compared (Figure 3.5). The result of the two contrasting topographic features demonstrates that the overall correlation of satellite rainfall estimates is better in lowland than in highland areas. A relative poorer performance in the mountainous area is notable for all products. This is in agreement with studies (Kimani et al., 2017; Worqlul et al., 2014; Dinku et al., 2007) who reported that satellite rainfall products have challenges to estimate orographic precipitation in basins with complex topography. Dinku et al. (2007) indicate that a very cold surface and warm orographic rains over mountainous area challenges to rainfall estimates by IR and PM algorithms. This suggests that the satellite products may not accurately capture the spatial pattern of seasonal rainfall in complex topographic areas like the Upper Tekeze basin. Further, the majority of products overestimate rainfall in the lowlands and underestimate rainfall in the highlands (Table A-2 and Figure A-1). However, comparing both topographic features, the magnitude of underestimation was greater than that of overestimation in most products. For example, CHIRPS, 3B42V7, and CMorph underestimated the wet season rainfall over the highland area by 32, 28, and 52%, while they overestimated by 18, 21, and 28% the lowland rainfall, respectively. It is also noted that CMorph has consistently overestimated the rainfall in all temporal scales during the point-to-pixel-comparison, whereas underestimated in the highland areas using an areal rainfall comparison approach. This discrepancy could be attributed to the effect of complex topography when the point rainfall from sparse and unevenly distributed gauges was converted into areal rainfall.

In summary, based on the comprehensive evaluation at different temporal and spatial scales, the CHIRPS outperformed the other products in all conditions. Next to CHIRPS, three products (RFEv2, 3B43V7, and PERSIANN) showed relatively good agreement with the observed rainfall. However, the ranks of these products were not consistent in all temporal scales. The better performance of CHIRPS can be explained by its high spatial resolution and by the fact that it considers topographic effects (Katsanos et al., 2015). Moreover, it performed better because it includes long-term climatic, CHG Precipitation Climatology (CPHClim) map (Mohamed et al., 2005). The good performance of 3B42V7 and RFEv2 is due to the fact that these products have a bias correction that is based on rain gauge data (Katsanos et al., 2015). GPCP and CMap products showed lower performance because of their coarse spatial and temporal resolution as well as that they suffer from inhomogeneity (Feidas, 2010; Xie and Arkin. 1995).

Our findings are in agreement with similar studies (e.g., studies such as Bayissa et al., 2017; Kimani et al., 2017; Dembélé and Zwart, 2016; Hessels, 2015, Katsanos et al., 2015; Dinku et al., 2007). Hessels (2015) compared 10 satellite products over the Nile basin and, CHIRPS and 3B42V7 were found to be the best-performing products. Bayissa et al. (2017) demonstrated that CHIRPS estimates performed better than PERSIANN and TARCAT over the Upper Blue Nile basin. Similarly, Gebremichael et al. (2014) and Dinku et al. (2007) showed that CMorph and 3B42V7 well performed in the rugged terrains of this neighbouring basin. CMorph, PERSIANN, and ARCv2 better captured the observed rainfall while CMap and GPCP poorly performed at all spatiotemporal scales. Dinku et al. (2007) found that CMap and GPCP products poorly performed compared to 3B42V7 and CMorph in Ethiopia.

The performance of all products consistently improved with ground measurements at larger timescales. This is due to counter balancing of variabilities when accumulated from shorter to longer timescales. Dembélé and Zwart (2016) and Guo and Liu (2016) showed that the performance of satellite estimates improved as time step increased. In contrast, performance did not uniformly improve increasing spatial scale. It increased from point to the areal/spatial comparison in the lowlands, whereas decreased in the highlands. Considering the complex topography and sparse rain gauge stations, the overall performance of CHIRPS and 3B42V7 products is very good and can be used for any hydrological and overall water management applications in the region.

3.4 CONCLUSIONS AND RECOMMENDATIONS

This study evaluated the performance of eight-satellite based rainfall products ranging from high to low resolution over the Upper Tekeze basin. These products were evaluated and compared with ground rainfall stations during 2002-2015. A comprehensive approach was applied that included point-to-pixel and areal averaged comparisons at different temporal scales (daily, monthly, and seasonal). The relationship between rainfall and elevation was also analysed based on the elevation category to identify the effects of topography on the performance of the products.

The results showed that the CHIRPS rainfall estimates outperformed other products in all time and space domains. This product achieved acceptable correlation coefficients (>0.5) and PBIAS, RMSE and MAE values for all conditions. Next to CHIRPS, estimates from RFEv2, 3B43V7 and PERSIANN showed a good agreement with the observed rainfall. However, the ranking of these products was not consistent for all temporal scales. These products performed better in the order of RFEv2, 3B43V7 and PERSIANN at the daily timescale, while this rank was changed to 3B43V7, RFEv2 and PERSIANN at monthly timescale. After CHIRPS, PERSIANN achieved a good agreement with the observed rainfall for the wet season comparison, followed by RFEv2 and 3B43V7, respectively. Compared to these products, CMorph and ARC2 achieved lower scores at most temporal and spatial scales. The performance of CMap and GPCP was poor over the various conditions. A relatively lower performance is notable for all products in the mountainous areas.

The agreement between the estimates of satellite products and rain gauge observations improved with an increase in timescale. This is most likely due to the fact that errors at smaller timescales offset each other when aggregated. All satellite estimates suffered from under- and over-estimation at different time and spatial scales. Underestimation dominated in the mountainous areas. 3B42V7 and CMorph consistently overestimated rainfall for the pixel-to-point rainfall comparison, and in the lowland for the areal comparison. These products underestimated the rainfall in the highland areas during the areal rainfall comparison approach. In contrast, all the remaining products underestimated the rainfall consistently at all spatiotemporal scales. CMap, ARC2 and GPCP underestimated rainfall the most in mountainous areas. Another key finding of this study is that unlike in time, the performance of the products at different spatial scales did not show a uniform pattern. The performance improved when increasing the areal averaged rainfall in the lowlands, whereas it decreased at larger spatial scale in the highlands. ARC2, CMorph, GPCP, PERSIANN and RFEv2 products better performed at the pixel-to-point rainfall comparison than using areal averaged rainfall at the whole basin, whilst CHIRPS, TRMM, and CMap performed better the other way round. Poor performance over the mountain areas contributed to a slightly lower performance of the products at larger spatial scales. Moreover, systematic errors during the interpolation of observed rainfall over the complex topography of the basin might have contributed to the overall lower performance at the basin scale.

The ranking of these rainfall products considered in this study may not be absolute as validation of these products in different study periods and conditions could result in different rankings. Interpolation of the sparse and unevenly distributed rain gauges over complex terrains may also introduce uncertainties and therefore limits the validity of the result. However, considering the current data availability, the result of this study provides a basis for the utilization of satellite rainfall estimates over the complex topography of the Tekeze basin, at least for monthly and seasonal timescales. It will be a useful reference for future applications of satellite rainfall, especially in rain gauge sparse and ungauged basins with rugged terrains.

Chapter 4

Temporal and Spatial Changes of Rainfall and Streamflow [2]

The Upper Tekeze river basin–part of the Nile basin, is characterized by high temporal and spatial variability of rainfall and streamflow. In spite of its importance for sustainable water use and food security, changing patterns of streamflow and their association with climate change is not well understood in the basin. Improving the understanding of the linkages between rainfall and streamflow trends and identifying possible drivers of streamflow variabilities is essential for water resources management. Trend analyses and change point detections of rainfall and streamflow were analysed using data records for 21 rainfall and 9 streamflow stations. The nature of changes and linkages between rainfall and streamflow were examined for monthly, seasonal and annual flows, as well as Indicators of Hydrological Alteration (IHA).

The result showed that rainfall did not show statistically significant changes. In contrast, trend analyses on the streamflow showed both significant increasing and decreasing patterns. A decreasing trend in the dry (October to February), short (March to May), main rainy seasons (June to September) and annual totals are dominant in the stations. Only one out of nine gauging stations experienced increasing flow significantly in the dry and short rainy seasons, attributed to the construction of Tekeze hydropower dam upstream this station in 2009. Overall, streamflow trends and change point timings were found to be inconsistent among the stations. Changes in streamflow without significant change in rainfall suggests factors other than rainfall drive the change. Most likely the observed changes in streamflow regimes could be due to changes in catchment characteristics of the basin. Further studies are needed to verify and quantify the hydrological changes shown in statistical tests by identifying the physical mechanisms behind those changes. The findings from this study are useful as a pre-requisite for studying the effects of catchment management dynamics on the hydrological variabilities in the basin.

[2] Based on Gebremicael, T.G., Mohamed, Y.A., van der Zaag, P., Hagos, E.Y., 2017. Temporal and spatial changes of rainfall and streamflow in the Upper Tekezē–Atbara river basin, Ethiopia. *Hydrol. Earth Sys. Sci.*, 21, 2127-2142.

4.1 INTRODUCTION

Recent changes in climatic conditions combined with other anthropogenic factors have increased the concern of the international community on water resources management in basins (Jones et al., 2015; Zhang et al., 2008). Understanding climate change and its impact on hydrological variability is important for water management, and thus has received attention from researchers in different parts of the world (e.g., Zhao et al., 2015; Zhan et al., 2014; Tekleab et al., 2013; Wang et al., 2013; Pano et al., 2010; Kim et al., 2008; Ma et al., 2008). These studies investigate how climate change reflected in changing rainfall patterns affects the hydrological regimes of river basins.

Identifying the trends and linkages between rainfall and streamflow is fundamental to understand the influence of climate change on the hydrological variability of a basin. Many studies (e.g., Zhao et al., 2015; IPCC, 2013; Shi et al., 2013; Tekleab et al., 2013; Tesemma et al., 2010) have shown that rainfall is the primary atmospheric factor that directly affects the streamflow patterns. The impact of climate change on hydrology varies from place to place. For example, Ma et al. (2008) for the arid region of northwest China, Zhang et al. (2011) for China, Zhao et al. (2015) for the Wei river basin of China, Love et al. (2010) for the Limpopo river of Southern Africa and Abeysingha et al. (2015) for the Gomti river basin in north India, found that a decreasing trend of rainfall resulted in a significant reduction in streamflow. In contrast, Masih et al. (2011) in the Zagros mountains of Iran, Wilk and Hughes (2002) in South India and Abdul Aziz and Burn (2006) in the Mavkenzie river basin of Canada, reported that a trend of increasing rainfall has significantly increased river flows. There are also a number of studies (e.g., Hannaford, 2015; Saraiva et al., 2015; Wang et al., 2015; Gebremicael et al., 2013; Tekleab et al., 2013) who found that changes in rainfall are not sufficient to explain the trends in the streamflow.

In Ethiopia, few studies have analysed the trend of hydro-climatic variables including, streamflow and rainfall. Conway and Hulme (1993) reported declining annual rainfall over the Blue Nile and Atbara basins resulting in a reduction of river flows between 1945 and 1984. In contrast, recent investigations by Tesemma et al. (2010), Tekleab et al. (2013) and Gebremicael et al. (2013) agreed that rainfall over the Upper Blue Nile basin did not show a statistically significant trend for the last 40 years (1964-2005). Despite that the pattern of rainfall remained constant, hydrological flows in the basin showed a heterogeneous trend. Rainy (June to September) and dry (October to February) season flows at the Upper Blue Nile basin outlet (El Diem) have significantly increased and decreased, respectively, while the mean annual runoff did not show a statistically significant trend. This implies that trends observed in the river flows may not be attributed to climate change but rather to changes in catchment characteristics. The lack of consensus in the literature may also show that there is still considerable uncertainty about the impact of climate change on the hydrological regimes of the region. The length of the statistical record has a direct implication on the results of the trend analyses and some of the observed discrepancies could be because of applying different periods of time series data. For example, Dixon et al. (2006) investigated the impact of record length on the trend pattern of streamflow in Wales and central England and their results indicated that trends over 50 to 60 years showed a statistically significant increasing trend, while for a medium record length (30–40 years) no such trend was detected. Meanwhile, record length of less than 25 years tended to show statistically significant increasing trends. This shows that trend analyses are sensitive to the time domain and careful attention should be given during analyses. Moreover, the variability in the climatic zone within a basin may also influence the hydrological regimes. Many studies (e.g., Li & Sivapalan, 2011; Castillo et al., 2003; Yair & Kossovsky, 2002) reported that the spatio-temporal runoff generation in semi-arid areas is strongly non uniform as runoff generation controlling factors are different from that of a humid environment.

With regard to the Upper Tekeze river basin, it lacks a comprehensive study of the hydro-climatic trends. Seleshi and Zanke (2004) attempted to investigate the pattern of rainfall over the upper part of Tekeze River basin by considering only one climatic station. Their output demonstrated that the amount of rainfall remained constant for the past 40 years (1962–2002). Despite the importance of streamflow to ensure sustainable water resource utilization and food security in the semi-arid regions of the country, long-term trends and change point of flow regimes and the association with climate change are not yet well understood. Therefore, it is important to understand the connections between rainfall and streamflow trends of the basin and establish whether hydrological variability is driven by changes in climate or by changes in catchment characteristics or both. This paper is intended to (i) investigate the spatiotemporal variability of rainfall and streamflow in the headwaters of Tekeze basin, (ii) identify any abrupt changes if significant trends exist, and (iii) explore the impact of climate change reflected in change in rainfall patterns on the hydrological variability of the basin.

4.2 DATA AND METHODS

Spatio-temporal datasets of rainfall and streamflow are required for the trend and change point analyses. These statistical analyses directly depend on the quality and length of the time series data. Therefore, much effort was given to verify the accuracy of the rainfall and streamflow data. These time series data are summarized in section 4.2.1 and 4.2.2.

4.2.1 Rainfall data

For this study, daily rainfall data since 1953 were used from 21 stations located within and surrounding the basin (Figure 4.1 and Table 4.1). These data were provided by the Ethiopian National Meteorological Service Agency. After scrutiny of all stations, only 21 out of more than 75 stations in the basin were considered for further analyses. The length of the data records varying from station to station, whilst all gauging stations with at least 30 years of continuous and relatively good quality of observed data were taken into account. A 30 years record period is a reasonable minimum length for applying statistical trend analyses of rainfall (Love et al.,

2010; Longobardi and Villani, 2009). The location and general information of all rainfall stations are shown in Figure 2.1 and Table 4.1.

Table 4.1: General information of rainfall stations, latitude and longitude, altitude (Alt.) in m.a.s.l, mean annual rainfall (mm/year), standard deviation (mm/year), and % age of missing data

Station name	Lat°	Long°	Alt. (m)	Recording period	Analyses period	Mean	SD	CV (%)	Missing data (%)
Mekelle	13.45	39.53	2,260	1952-2015	1953-2015	576	141	24	0.0
Mychew	12.69	39.54	2,432	1953-2015	1953-2015	697	158	23	6.3
Axum	14.12	38.74	2,200	1962-2015	1963-2015	690	159	23	9.4
Gonder	12.60	37.50	2,316	1952-2015	1964-2015	1090	195	18	0.0
Adwa	14.16	38.90	1,950	1964-2015	1967-2015	705	176	24	7.6
Mykinetal	13.94	38.99	1,815	1967-2015	1967-2015	585	129	22	8.1
Shire	14.10	38.28	1,920	1963-2015	1968-2015	953	203	21	10.1
Adigrat	14.00	39.27	2,470	1970-2015	1970-2015	596	172	29	2.1
Adigudem	13.16	39.13	2,100	1975-2015	1971-2015	498	156	31	2.2
E/hamus	14.18	39.56	2,700	1971-2015	1971-2015	651	214	33	2.2
Hawzen	13.98	39.43	2,255	1971-2015	1971-2015	505	116	23	6.6
Illala	13.52	39.50	2,000	1975-2015	1975-2015	563	138	25	4.9
H/Selam	13.65	39.17	2,630	1973-2015	1973-2015	685	168	24	0.0
AbiAdi	13.62	39.02	1,850	1961-2015	1973-2015	861	246	29	2.3
Samre	13.13	39.13	1,920	1967-2015	1978-2015	650	188	29	6.1
D/tabor	11.85	38.00	2,969	1974-2015	1974-2015	1502	264	17	2.3
Dengolat	13.19	39.21	1,950	1975-2015	1975-2015	617	166	27	2.4
Lalibela	12.03	39.05	2,450	1972-2015	1978-2015	789	169	21	5.3
Wukro	13.79	39.60	1,995	1962-2015	1985-2015	485	139	29	9.4
Kulmesk	11.93	39.20	2,360	1973-2015	1985-2015	668	180	27	3.2
Debarik	13.15	37.90	2,850	1955-2015	1984-2015	1104	231	21	6.2

Visual inspection, linear and multiple regression analyses between neighbouring stations and other global datasets, including New_LocClime software package (Grieser et al., 2010), CHIRPS (Funk et al., 2014) and TRMM (Simpson et al., 1988), were applied for data analyses and validation, detecting outliers, filling missing values and reliability checking for all gauging stations. Both CHIRPS and TRMM (3B42v7) satellite products were the best performed products over the basin (Chapter 3). The rainfall datasets were found to be reliable to be used for statistical analysis.

The coefficient of variation in annual rainfall of the basin ranges from 18 % in the southern to 33 % in the eastern and northern parts of the basin. As presented in Table 4.1, all but two stations have a coefficient of variation below 30 % which is acceptable limit for data validation (Sushant et al., 2015; Medvigy and Beauliew, 2011). To ensure data continuity and integrity, missing rainfall data of less than 1 year were estimated from global and neighbouring stations and data gaps larger than 1 year were excluded from the analyses. Based on these data screening and analyses methods, rainfall data with missing values less than 10 % have been used in the analyses (Table 4.1).

4.2.2 Streamflow data

Streamflow data from all gauging stations in the basin were obtained from the Ethiopian Ministry of Water Resources and Energy. Although the recording of flow data over the basin started in the late 1960s, it was discontinued for most of the gauging stations during the civil war in the 1980s. To this effect, only 9 out of the total 39 stations have an extended period of more than 20 years data and these were used in the analyses (Table 4.2). The location and general information of all flow stations of the basin are summarized in Figure 2.1 and Table 2.1.

Spatio-temporal trends analysis can be affected by the chosen length of records. A longer period of historical data increases the visibility of dominant trends and the reliability of results from trend analyses, while a shorter period enhances spatial coverage of streamflow by including more representing stations at different parts of the basin. To better account for the spatial variability of streamflow, a length of more than 20 years data is desirable for trend analyses of streamflow (Abeysingha et al., 2015; Saraiva et al., 2015; Abdul Aziz and Burn, 2006).The sensitivity of trend to the length of flow record is discussed in the introduction (Section 4.1). The average annual flows of each station indicate that hydrological responses are spatially uneven over the basin (Table 4.2). For example, Illala and Werie have higher streamflow per unit area as compared to Genfel and Geba 2 (near Adikumsi) tributaries. Despite large difference in the drainage area (Table 4.2 and Figure 2.1), approximately, the same volume of runoff is contributed to the Geba at Adikumsi from Genfel and Illala tributaries. Moreover, looking into the drainage area of Geba 1(near Mekelle) (4,590 km^2) and Werie (1,770 km^2), more water is discharged from Werie (5 %) than from Geba at Mekelle (6 %) to the basin outlet (at Tekeze 2) which suggest a high variability in hydrological response to catchment characteristics. Furthermore, the coefficient of variation (CV) and standard deviation (SD) shown in Table 4.2 also indicates that there is high inter-annual variability of streamflow in all monitoring stations

Table 4.2: Summary of statistical analysis on the hydrological flows considered for further analysis

Station name	Analysis period	Catchment area (km²)	Annual average flow (m³s⁻¹)	mm/year	SD -	CV (%)	Missing data (%)
Siluh	1973-2015	967	1.0	33	0.28	28	4.6
Illala	1980-2015	341	0.6	55	0.192	32	2.8
Genfel	1992-2016	733	0.6	26	0.246	41	2.1
Werie	1994-2015	1,770	10.1	180	2.828	28	0.8
Agula	1992-2016	692	1.1	50	0.374	34	0.3
Geba 1	1991-2015	2,445	3.9	50	1.131	29	4.0
Geba 2	1994-2015	4,590	14.2	98	2.982	21	0.0
Tekeze 1	1994-2015	1,002	3.0	94	0.69	23	3.2
Tekeze 2	1994-2015	45,694	219.5	151	30.73	14	0.0

As hydro-meteorological data in the basin, if not in all basins in Ethiopia, is very limited and with many gaps, it is critical to carefully screen and check their quality before using them for analyses. Hence, the raw data were visually inspected and screened for typos and outliers. Each station was carefully checked for data consistency by comparing to the nearby, upstream and downstream stations. Relationships between neighbouring stations can give a preliminary evidence on the reliability of time series data provided that there is no man-made water storage above the station (Hong et al., 2009). For example Figure 4.1 compares the consistency of annual flows of three upper stations (Siluh, Genfel and Agula) against their combined flow measured in the downstream at Geba 2 gauging stations of the Geba sub-catchment. These discharge single mass curve clearly indicates that the streamflow in all stations (upstream, neighbouring) have similar trends. Identified unreliable data such as outliers and missing values were fixed by removing outliers or large data gaps and interpolating from upstream/downstream stations after comparing its upper and lower boundary limits. Furthermore, heterogeneity of the time series data was also assessed using the double mass curve and residual mass plot methods. Example of such mass curves are given as supplementary file in Appendix B (Figure B-1).

The monthly hydrological flow data were aggregated from the daily data and the seasonal and annual data was calculated from the monthly data. In order to remove unreliable data whilst including more stations to increase spatial coverage, missing data for more than two years were excluded from the analyses. However, during the peak rainy season missing data for more than two weeks were excluded from the analyses. As presented in Table 2.1, only data with missing values of less than 5% were considered for further analysis. The reason to exclude only two weeks was to minimize untrustworthy data as more than 80 % of the river flow is generating during only two months (July and August).

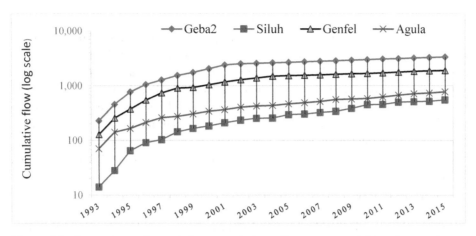

Figure 4.1. Comparison of cumulative annual streamflow among neighbouring, upstream and downstream stations in Geba sub-catchment

4.2.3 Trend analyses method

To identify the trends in rainfall and streamflow, a non-parametric Mann-Kendall (Kendall, 1975) statistical test is applied. The Mann–Kendall test (MK), is a rank based method that has been widely used to detect the trend of hydro-climatic time series data in different parts of the world (e.g. Jones et al., 2015; Návar, 2015; Wang et al., 2015; Mohamed & Savenije, 2014; Gebremicael et al., 2013; Tekleab et al., 2013; Abdul Aziz and Burn, 2006;). The procedure of MK testing starts by calculating the MK statistics using Eq. (4.1) (Yue et al., 2002).

$$S = \sum_{i=1}^{n=1} \sum_{j=i+1}^{n} sgn(x_j - x_i), \quad \text{where } sgn(\theta) = \begin{cases} +1 \\ 0 \\ -1 \end{cases} \text{ if } \begin{bmatrix} \theta > 0 \\ \theta = 0 \\ \theta < 0 \end{bmatrix} \tag{4.1}$$

where x_j and x_i are the data values in time j and i and j>i, respectively, and n is the length of data set. The normalized test statistics Z of MK test and the variance VAR(S) were calculated as shown in Eq. (4.2) and Eq. (4.3).

$$Z = \begin{cases} \frac{S-1}{\sqrt{V(S)}} \\ 0 \\ \frac{S+1}{\sqrt{V(S)}} \end{cases} \text{ if } \begin{bmatrix} S > 0 \\ S = 0 \\ S < 0 \end{bmatrix} \tag{4.2}$$

$$V(S) = \frac{1}{18}[n(n-1)(2n+5)] \tag{4.3}$$

where s and v(S) are the Kendall's statistics and Variance, respectively.

The MK test calculates Kendall's statistics s, which is the sum of the difference between data points and a measure of associations between two samples (Kendall's tau). The MK test, accepts the null hypothesis if $-Z \leq Z_{cr} \leq Z$, where Z_{cr} is critical value of the normalized statistics Z at 5 % confidence level (1.96). Positive and negative values of those parameters (z, s, and tau) indicate an 'upward trend' and 'downward trend', respectively. In order to evaluate the trend results, the Z value combined with the computed two-tailed probability (P) were compared with the user defined confidence level (5 %) of the standard normal distribution curve. The MK test is commonly used and suitable to identify trends in water resources as it is not affected by the distribution, outliers and missing values of time series data (Zhang et al, 2011, 2008; Yue et al., 2003, 2002).

The existence of serial correlation in the time series data may affect trend detection in the non-parametric trend test methods (Masih et al., 2011; Zhang et al., 2011). The Trend-Free Pre-Whitening (TFPW) method (Yue et al., 2003) was employed to avoid serial correlations in the data. This method is found to be the most powerful tool to remove a serial correlation time series if it exists (Mohamed & Savenije, 2014; Tekleab et al., 2010; Burn et al., 2004; Yue et al., 2003). The data series was pre-whitened using the formula shown in Eq. (4.4).

$$Y_1 = Y_t - r1.Y_{t-1} \tag{4.4}$$

where r1 is the estimated serial correlation coefficient, Yt is trended series for time interval t, and Y1 is data series without auto-regressive, and Yt-1 is the original time series value. Detailed descriptions of TFPW can be found in literature (e.g., (Gao et al., 2011; Tekleab et al., 2010; Burn et al., 2004; Yue et al., 2003). Finally, the MK test was applied to the TFPW data series for analysing the gradual change in the rainfall and streamflow.

4.2.4 Change point detection method

To estimate the occurrence of an abrupt change, a non-parametric Pettitt test (Pettitt, 1979) was applied to the TFPW data series. The Pettitt test is a rank-based and distribution-free test for identifying if there is a significant change between cumulative functions before and after a time instant. The Pettitt test considers a sequence of random variables X1, X2, ..., XT that can have a change point at τ if Xt for t, ..., τ have common distribution function f1(x) and Xt for t= τ+1, ..., T have a common distribution function f2(x), and f1(x) ≠ f2(x). The test statistics KT= Max │Ut, T│, $1 \leq \tau < T$ and associated probability (P) used to test were computed using equations found in Gao et al. (2011). The test was evaluated against a user-defined significance level (5 %) and P values less than 5% were considered as a statistically significant change in the data series. The 5% significance level was chosen as it is commonly used in the hydro-climatic trend analyses (e.g. Tekleab et al., 2013; Gebremicael et al., 2013; Saraiva Okello et al., 2015). This Pettitt technique has been widely used to detect time change points in the hydro-climatic data (e.g., Gebremicael et al., 2013; Tekleab et al., 2013; Gao et al., 2011; Zhang et al., 2011; Love et al., 2010; Ma et al., 2008).

4.2.5 Hydrological alteration indicators

Temporal and spatial streamflow variability can also be characterized and quantified using hydrologic alteration indicators. The Indicators of Hydrologic Alteration (IHA) software developed by the US Nature Conservancy (Mathews and Richter, 2007) were applied to assess the degree of hydrological alteration. Eight parameters out of the total 33 IHA parameters were selected for this study. The selected parameters are magnitude and duration of annual extreme water conditions (e.g. 1–day, 3–day, 7–day minimum and maximum flows), rate and frequency of water condition changes (e.g. rising rate and falling rate) and magnitude of monthly water conditions (e.g. monthly flows). Such IHA parameters are common in characterizing of hydrological regimes influenced by climate and anthropogenic factors (Saraiva et al., 2015; Masih et al., 2011; Mathews and Richter, 2007; Tayler et al., 2003). The consistency of those parameters was analysed and compared with the user-defined P values (5 %).

4.3 RESULTS AND DISCUSSION

According to NMSA (1996), major seasons in the study area are the rainy (June-September), dry (October-February) and short rainy seasons (March-May). Before detecting trends in precipitation and hydrological flows, serial correlation existence in all datasets was tested at monthly, seasonal and annual scales. Accordingly, 9.1 – 18 % of the monthly, seasonal and annual data of the rainfall stations were found to have a statistically significant auto–correlation at 95% confidence level. This indicates that a false significant trend might have appeared in some of the stations when actually there is no trend because of auto-correlation in the data series. Similarly, 3.2 – 6 % of the monthly and seasonal flows showed statistically significant auto-correlation. It is unclear why the monthly and seasonal fluxes showed stronger autocorrelation than annual. The probable reason could be because of storage properties in the catchments, unreliable data and missing values (Abeysingha et al., 2015; Hirsch and Slake, 19). Furthermore, continuous constant observations in the dry months, where river discharges are very low may have increased the degree of similarity among consecutive observations. To avoid such spurious trend detection, serial correlation problems in all periods were eliminated using TFPW techniques before trend analyses. Comparison of rainfall and streamflow before and after TFPW at different scales (monthly, seasonal and annual) is presented in Appendix B (Figure B-2 and Figure B-3, respectively).

4.3.1 Rainfall variability over the Upper Tekeze basin

The presence of monotonic increasing/decreasing trends in monthly, seasonal and annual rainfall of 21 gauging stations was tested using the MK test. The results for seasonal and annual rainfall are summarized in Table 4.3. Positive and negative values of Z statistics show increasing and decreasing trends, respectively. Z statistics in bold illustrate statistically significant trends of rainfall. The spatial distribution of observed significant and non-significant trends of annual rainfall over the basin is also given in Figure 4.2.

Results of the trend analyses were used to identify if the time series of annual and seasonal rainfall had a statistically significant trend in the last 30–60 years (Table 4.3). Figure 4.2 shows the spatial variability of rainfall on annual scale throughout the basin. Except for two stations (Axum and Shire), there is no significant in the rainfall trends of the basin. Both Axum and Shire stations which are located in the North West part of the basin showed an increasing and decreasing trend, respectively (Figure 4.2). The possible reason for obtaining the different result in these stations could be because of unreliable data as both stations have the highest percentage of missing data compared to the remaining stations. However, although statistically not significant, statistical indices of the test revealed a tendency of decreasing rainfall patterns in the eastern and northern part of the basin during the main rainy season and the entire year (Table 4.3). Meanwhile, there is an increasing tendency in the southern and western parts of the basin for the same time scales. With regard to monthly rainfall, despite there was some temporal and spatial variability, no dominant trends are found in the majority of the months (Table B-1).

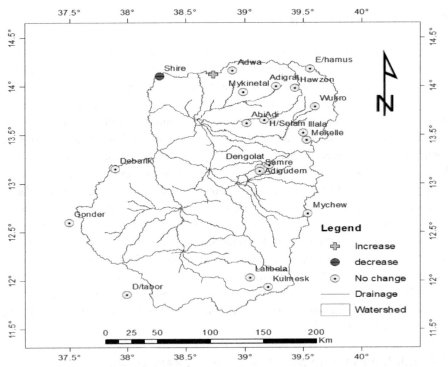

Figure 4.2: Location of rainfall stations with increasing (+), decreasing (-) and no change (0) trends on annual rainfall

The Pettitt test was used to identify if there is a breakpoint in the data series. Similar to the MK test, the majority of rainfall stations did not show statistically significant change points at 5 % significance level. For example, the change point for annual and seasonal rainfall of some stations can be seen in Figure 4.3. Figure 4.3a and 4.3d illustrate that annual rainfall in Mekelle (AP) and Adigrat stations do not show an upward or downward shift in the given time domain. An increasing and decreasing change point of annual and short rainy season (March–May) rainfall in Axum and AbiAdi stations are shown in Figure 4.3b and 4.3c, respectively.

In general, most of the rainfall stations across the basin did not experience a trend at 5 % significance level. The result also reveals that even though there is no dominant trend, monthly rainfall over the basin is observed to be more variable compared to the seasonal and annual rainfall (Table B-1). The possible reason could be the amount of rainfall in a given month is shifted to the next or previous month and the station might be receiving the same amount of seasonal rainfall but varying in distribution among the months. Another possible reason could be a topographically induced climate. For example, Bizuneh (2013) noticed that monthly rainfall variability in Siluh catchment of Geba basin is dependent on altitude. Furthermore, observed monthly rainfall variability might be due to large-scale oscillation (e.g. ITCZ) variability rather than long–term climate variability in the basin.

Table 4.3: Summary results Z statistics and P-values (in brackets) on seasonal and annual rainfall trends. Negative/positive Z value indicates a decreasing/increasing trend and in bold a statistical significant trend at 5%.

Station name	Rainy season Jun to Sep	Dry season Oct to Feb	Short rainy Mar to May	Annual Jan to Dec
Mekelle (AP)	-1.5 (0.10)	1.7 (0.21)	-1.2 (0.52)	-1.8 (0.14)
Mychew	-1.1(0.48)	1.0 (0.25)	0.9 (0.48)	-0.7 (0.68)
Axum	1.5 (0.23)	0.3 (0.81)	0.3 (0.81)	**2.5 (0.04)**
Gonder	1.1 (0.66)	-1.7 (0.23)	1.8(0.25)	0.8 (0.73)
Adwa	**4.6 (0.02)**	1.6 (0.53)	1.1 (0.42)	0.8 (0.06)
Mykinetal	1.7 (0.47)	-1.4 (0.40)	-1.8 (0.08)	-0.3 (0.90)
Shire	1.6 (0.06)	1.4 (0.20)	1.5 (0.18)	**-2.5 (0.02)**
Adigrat	-0.4 (0.73)	1.1 (0.37)	1.1 (0.39)	-0.1 (0.96)
Adigudem	0.0 (0.93)	0.1 (0.92)	-0.8 (0.52)	-0.3 (0.78)
Edagahamus	-0.6 (0.59)	-1.2 (0.27)	-1.7 (0.11)	-1.1 (0.33)
Hawzen	0.7 (0.57)	-1.4 (0.05)	-1.1 (0.36)	-0.4 (0.72)
Illala	1.5 (0.13)	0.3 (0.72)	1.6 (0.09)	1.6 (0.07)
Hagereselam	0.2 (0.85)	-0.3 (0.83)	-1.7 (0.16)	-1.0 (0.47)
AbiAdi	0.9 (0.43)	-1.5 (0.06)	**2.9 (0.02)**	1.9 (0.06)
Debretabour	1.3 (0.06)	1.5 (0.04)	0.1 (0.90)	1.9 (0.06)
Dengolat	-0.6 (0.57)	0.1 (0.94)	1.7 (0.09)	1.9 (0.05)
Lalibela	0.8 (0.20)	-0.6 (0.38)	-1.1 (0.08)	-1.1 (0.07)
Wukro	-1.4 (0.31)	0.8 (0.57)	1.2 (0.52)	1.6 (0.08)
Kulmesk	0.5 (0.06)	-0.3 (0.13)	0.1 (0.83)	0.1 (0.67)
Debark	0.3 (0.36)	0.4 (0.36)	1.8 (0.07)	1.7 (0.06)
Samre	-0.3 (0.84)	-1.9 (0.09)	-0.3 (0.79)	-0.3 (0.79)

The results of this study are consistent with Seleshi and Zanke (2004) who found no significant trends of rainfall at Mekelle station. Results from neighbouring catchments of similar climate characteristics and applying the same methods of trend analyses have also shown that the pattern of rainfall remained constant for the last 40 years which is in agreement with our finding (e.g., Gebremicael et al., 2013; Tekleab et al., 2013).

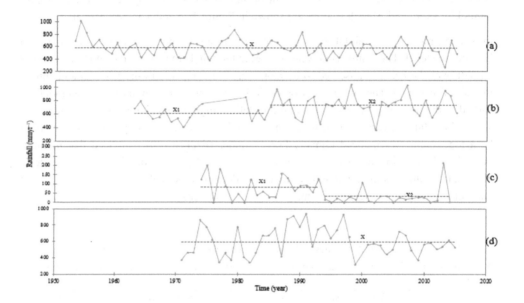

Figure 4.3: Pettitt homogeneity test of selected rainfall stations (a) annual rainfall in Mekelle (AP), (b) annual rainfall in Axum, (c) short rainy season in AbiAdi, (d) annual rainfall in Adigrat. X1 and X2 are average values of rainfall before and after change point.

4.3.2 Streamflow variability in the basin

Streamflow of nine gauging stations (Table 4.2) was analysed for long term trend detection using MK and Pettitt test. Table 4.4 summarizes the results from the MK test and positive and negative values of Z statistics associated with the computed probability (P–value) shows an increasing or decreasing trend. Statistically significant trends are shown in bold. Furthermore, the observed trends are also presented in Figure 4.4 to show the spatial variability of streamflow over the basin. Annual flow patterns exhibited a declining trend in the majority of stations although this time is more pronounced in the eastern part of the basin (Table 4.4). The change is found to be statistically significant at Siluh, Genfel, Geba stations. Interestingly, although there is a dominant decreasing pattern in the majority of stations, the annual flow at Embamadre (Tekeze 2) station did not significantly change. Seasonal streamflow of the stations was also analysed to further scrutinize temporal and spatial variability.

The analyses found that a significant decreas in dry season flow has occurred in most stations (Table 4.4). During the short rainy season, a decreasing trend has occurred in five stations, some of these trends being statistically significant. Nevertheless, in the dry and short rainy seasons flow has significantly increased only at the basin outlet (Tekeze 2). Streamflow showed a significant increasing trend at Tekeze 2 and a non-significant increasing trend at Tekeze 1 during the main rainy season. In contrast, a gradual decreasing pattern of flow was recorded in the remaining gauging stations with this change being significant in four stations (Table 4.4).

Table 4.4: Summary results of MK, Z statistics on streamflow trends. Negative/positive Z value indicates a decreasing/increasing trend and in bold a statistical significant trend at 5 % confidence level (Z= ±1.96).

Period	Siluh	Genfel	Agulae	Illala	Werie	Geba 1	Geba 2	Tekeze 1	Tekeze 2
Record length (year)	1973-2015 43	1992-2015 23	1992-2015 24	1980-2015 36	1994-2015 22	1991-2015 25	1994-2015 21	1994-2015 21	1994-2015 22
Annual	**-4.5**	**-3.1**	-0.9	-0.7	-1.1	**-2.1**	-1.7	0.6	1.0
Rainy season	**-3.1**	**-3.2**	-0.4	0.4	0.2	**-2.1**	**-2.8**	1.0	0.1
Short rainy	**-2.4**	**-3.1**	**-3.5**	**-3.1**	1.1	-1.5	**2.7**	0.63	**3.9**
Dry season	**-5.0**	**-3.0**	0.3	**-3.0**	-0.9	**-2.2**	**-3.3**	-0.2	**3.4**
Jan	**-5.1**	**-2.3**	-0.7	**-2.1**	1.4	**-2.7**	**-3.1**	0.4	1.6
Feb	**-4.5**	**-2.6**	-0.2	-1.4	0.8	**-2.8**	-0.4	-0.0	0.8
Mar	**-5.9**	-1.5	-0.5	-1.6	1.2	**-3.3**	-0.2	0.4	1.3
Apr	**-4.0**	**-2.4**	**-4.0**	**-2.9**	0.9	**-2.1**	**-2.6**	0.7	**2.4**
May	**-4.8**	**-2.1**	**-2.6**	**-2.2**	1.3	**-2.1**	-0.3	0.7	**2.3**
Jun	**-1.4**	-1.9	-0.9	-1.6	0.6	-1.5	-0.2	0.6	1.2
Jul	**-4.7**	**-2.6**	-0.7	-1.0	-0.5	-1.6	-0.4	-0.3	-0.1
Aug	-0.9	**-2.2**	0.8	1.0	-1.2	**-2.0**	**-3.4**	0.3	-1.1
Sep	1.1	-1.6	-0.3	1.5	-0.1	-1.4	-0.4	0.6	0.3
Oct	**-6.7**	-1.8	-0.3	-0.8	**3.8**	**-2.4**	-0.6	0.3	0.9
Nov	**-4.7**	**-4.8**	-1.2	-1.5	0.7	-1.5	-1.1	0.3	0.7
Dec	**-4.7**	**-2.6**	-0.2	-1.5	1.2	**-2.4**	**-3.4**	0.2	0.4

Geba1 = Geba near Adikumsi, Geba2 = Geba near Mekelle, Tekeze1 = Tekeze at Kulmesk, Tekeze2 = Tekeze at Embamadre

Majority of the gauging stations did not show a consistent trend in monthly streamflow. For example, discharge in Siluh and Genfel stations is characterized by a decreasing trend in most months. A significant decreasing trend is found in April and May flows of Agulae and Illala watersheds while the remaining months observed a decreasing trend that was not statistically significant (Table 4.4). The combination of Siluh, Genfel and Agulae tributaries at Geba station near Mekelle showed a decreasing trend in all months. Monthly flow patterns of the Upper Tekeze River Basin, the sum of all gauged and ungagged tributaries at Embamadre station, revealed a significantly increasing trend in April and May while all other months remained unchanged.

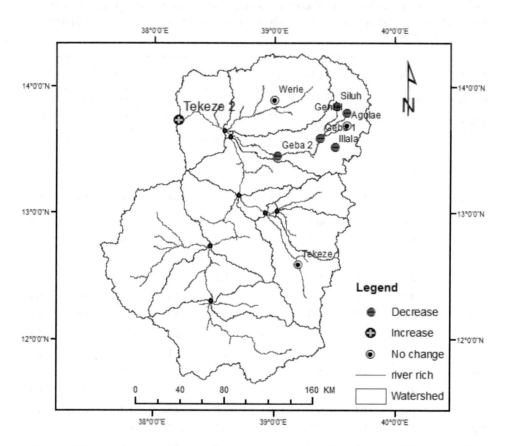

Figure 4.4: Location of streamflow stations with increasing (+), decreasing (-) and no change (0) trends on dry season flows

The Pettitt test was also applied to identify an abrupt change of streamflow in the catchments. The change points of annual and seasonal flow for selected stations are shown in Figure 4.5. For the annual flow, stations did not show consistent shifts across all stations. For example, annual flows in Siluh and Geba catchments shifted downward after 1992 and 2002, respectively, while no significant abrupt change was observed in Genfel and Tekeze 2 at Embamadre despite strong monthly and seasonal variability. Change points of seasonal flow for the same stations confirmed an abrupt change in the downward and upward directions (Figure 4.5). Dry and short rainy season flows in all stations except at the basin outlet showed a significant downward shift since the early 2000s. Conversely, an abrupt increase of streamflow has occurred at the basin outlet for the same seasons. The Pettitt test has also shown that hydrological flows during the rainy season remained constant in most stations (Figure 4.5).

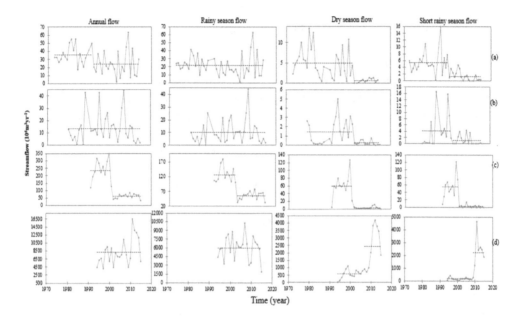

Figure 4.5: Annual and seasonal streamflow abrupt changes as determined by Pettitt test at (a) Siluh, (b) Illala, (c) Geba 1 and, (d) Tekeze 2

In summary, most stations exhibited a statistically significant change during the short rainy (5 stations) and dry seasons (6 stations). Similarly, trends in the main rainy season and annual flow showed a significant change in 3 to 4 of the stations. Several stations exhibited a change point of monthly streamflow (see Figure B-4). Both upward and downward shifts in streamflow were observed in many stations for the months of Jan, April, May and August. However, there was no dominant (increasing or decreasing) trend across the basin. For example, a change points of monthly (Jan, April, May and August) streamflow is observed in Siluh catchment (Figure B-4). A downward shift of monthly streamflow has occurred since 1996. The monthly stream flow of Genfel and Geba catchments has significantly declined starting from 2003 and 2004, respectively. In contrast, an upward shift of monthly streamflow was observed at Emabamadre for January, April and May which became significant after 2009.

Interestingly, some of the trends in the upper catchments counterbalance each other when combined in the downstream stations. For example, negative trends during the short rainy season in Siluh, Genfel, and Agulae cancelled out when combined at Geba near Mekelle. A remarkable result was that, although majority stations in the upper catchments showed a declining pattern of streamflow, the entire basin flows at the outlet did not show a negative trend. The significant increasing trend during the dry and short rainy season at this station is most likely due to the construction (2009) of Tekeze hydropower dam located 83 km upstream of the station. Change in catchment response from ungauged catchments of the basin might also have contributed to increasing the flow at this station.

The above results are in agreement with previous local studies (e.g. Abraha, 2014; Bizuneh, 2013; Zenebe, 2009) who found strong variability of stream flow in different sub-catchments of the basin. Compared to the neighbouring basin (Upper Blue Nile) studies by Gebremicael et al. (2013) at four stations and Tessema et al. (2010) at three stations who found a significant increasing trend of streamflow in short rainy, main rainy, annual flows and a decreasing trend in the dry season flow, the Tekeze basin, particularly in the semi-arid parts of the basin experienced a significant decreasing trend and high variability of streamflow. This variability is expected as land degradation (e.g. deforestation, over cultivation and grazing) in Tekeze basin is more pronounced than other basins in the country (Gebreyohannes et al., 2013; Gebrehiwot et al., 2011; Zenebe, 2009; Awulachew et al., 2007; Yazew, 2005).

4.3.3 Analysis of hydrological variability using IHA method

Although the previous analyses showed the long-term trends of rainfall and streamflow in the Upper Tekeze basin, it could not address the short period fluctuations of the hydrology within the catchment, and whether it can explain some of the results given above. The magnitude and duration of annual extreme conditions were also analysed using six IHA parameters (1–day, 3–day and 7–day annual minima and maxima). Results from these extreme conditions indicate a dominant significant decreasing trend in both minimum and maximum daily flows (Table 4.5). On the other hand, a significant increasing trend of the minimum flow is detected at Tekeze 2 (Embamadre) station.

The trends in the rate and frequency of changes in hydrological conditions were also explored using rise and fall rate parameters. Accordingly, the rising rate of daily flow of all stations remained constant while the daily falling rate has significantly increased in the tributaries and decreased in the basin outlet. It can be seen that the trend of minimum flows described by 1–day, 3–day and 7–day, is consistent with the trend of monthly and seasonal flows. Moreover, the IHA change point analyses have also shown shifts in minimum and maximum flow during the dry and wet seasons of the catchments. The extreme 1–day and 7–day minimum and maximum flows significantly shifted downward at the Siluh and Geba catchments. In the Tekeze at Embamadre station, the 1–day and 7–day minimum flows significantly increased around 2003, but 1–day and 7–day maximum flows remained unchanged. Illala catchment experienced a decrease of the 1–day minimum and an increase in the 1–day maximum annual flows with change points at around 2000 and 1995, respectively. Extreme high flows characterized by the 1–day and 7–day annual and maximum flows did not significantly change at the basin outlet which may be also due to homogenisation of the low flow and peak flow hydrograph after the construction of Tekeze hydropower dam above the station.

Table 4.5: Summary results of MK, Z statistics on IHA trends. Negative/positive Z value indicates a decreasing/increasing trend and in bold a statistical significant trend at 5 % confidence level (Z= ±1.96).

IHA parameters	Siluh	Genfel	Agulae	Illala	Werie	Geba 1	Geba 2	Tekeze 1	Tekeze 2
record	1973-2015	1992-2015	1992-2015	1980-2015	1994-2015	1990-2015	1994-2015	1994-2015	1994-2015
length (yr)	43	23	24	36	22	25	21	21	22
1-day minimum	**-4.3**	**-3.2**	-0.7	**-3.4**	0.8	**-3.6**	**-3.8**	-0.6	**2.8**
3- day minimum	**-3.5**	**-4.2**	-0.8	**-3.6**	0.8	**-3.4**	**-3.8**	-0.5	**2.2**
7-day minimum	**-3.5**	**-3.5**	-1.3	**-3.1**	0.7	**-3.9**	**-3.5**	-0.6	**2.1**
1-day maximum	**-5.7**	-1.0	-0.1	0.7	-1.3	**-2.7**	-1.0	0.8	-1.0
3-day maximum	**-5.5**	-1.3	-0.7	0.7	-0.9	**-2.6**	-0.8	-0.1	-1.5
7-day maximum	**-6.4**	**-2.2**	-0.4	0.1	-1.3	**-2.4**	-0.7	-0.6	-1.2
Rise rate	-1.1	-0.4	-1.3	1.0	1.2	-1.5	-1.3	1.0	-0.3
Fall rate	0.8	2.3	3.9	1.6	1.4	2.8	0.8	-2.3	-3.8

4.3.4 Drivers for streamflow variabilities

Climatic conditions, and in particular rainfall, as well as human activities in a catchment are the most important factors influencing the hydrological variability of streams. In this study, temporal and spatial analyses of rainfall from both MK and Pettitt tests showed that rainfall over the basin did not significantly change during the period of analyses. Streamflow, in contrast, did exhibit a strong temporal and spatial variability in the basin. This suggests that the change in hydrological flow is not significantly influenced by rainfall. The timing of observed trends in streamflow is not uniform, however, this may indicate that the impact of human interference and physiographic characteristics differ from sub-catchment to sub-catchment. Trend analyses is sensitive to the time domain as different results can be obtained for different periods. In this study, however, change points occurred at different times in most of the sub-catchments even for the same time domain (e.g. Genfel and Agula). This implies that the effect of changes in the underlying surface characteristics could be the physical mechanism behind those variations. Human interventions expressed in terms of water abstraction, implementation of large-scale soil and water conservation, deforestation, and afforestation in the upstream catchments are the more likely driving forces of changes in the flow regimes than climatic conditions. For example, Nyssen et al. (2004) and Belay et al. (2014) reported that a strong decrease of forest and bushland has occurred in favour of cultivable and grazing lands from the 1960s to early 1990s. However, the rate of deforestation and land degradation in most parts of the basin has slowed down and even started to recover by increasing the coverage of vegetation from the late 1990s onwards (Nyssen et al., 2010).

Increasing water abstractions, particularly in the semi-arid catchments of the basin, might have caused the decline of streamflow during dry and small rainy seasons. Several studies, e.g., (Kifle, 2015; Nyssen et al., 2010; Alemayehu et al., 2009) have shown that surface and shallow groundwater development and abstraction for irrigation have significantly increased since the mid-2000s, after implementation of intensive catchment management programmes. Moreover, a strong monotonic trend in streamflow without a significant change in rainfall during the rainy season could be attributed to the large-scale soil and water conservation interventions in the upstream watersheds. For example, Nyssen et al. (2010) and Abraha (2014) reported that integrated catchment management and land use change have significantly reduced streamflow in Geba catchment. Abraha (2014) showed the conversion of vegetation cover into agricultural land has increased runoff by 72% and decreased dry season flow by 32% over 1972-2003. Studies in neighbouring basins (e.g. Upper Blue Nile) also confirmed that conversion of vegetation cover into agriculture and bare land has caused an increase of surface runoff and decreasing of base flow up to 75% and 50%, respectively. On the other hand, soil and water conservation interventions have significantly increased the availability of groundwater at smaller watershed levels (Alemayehu et al., 2009; Negusse et al., 2013). All these studies are consistent with our findings that observed streamflow alterations in the basin are most likely the result of upper catchments interventions than changing patterns of rainfall. Quantifying the impacts of such factors at large scale is beyond the scope of this study and further investigations should be conducted to study the effect of anthropogenic factors on streamflow variability and change at different scales.

4.4 CONCLUSION AND RECOMMENDATIONS

This study presents a detailed statistical analysis on the existence of trends and point changes of rainfall and streamflow in the Upper Tekeze River basin. The analyses were carried out for 21 rainfall and 9 streamflow monitoring stations. Those stations were selected based on the availability and quality of data from 39 streamflow and more than 70 rainfall stations available in the basin. Linkages between the trends in rainfall and streamflow across the whole basin were carefully examined at different scales. Following these analyses the main driving force for streamflow variability over the basin is deducted.

Rainfall over the basin has remained constant in the last four decades. The 19 out of the 21 tested stations experienced neither increasing nor decreasing trends during the dry, short rainy, main rainy seasons and annuals at 95 % confidence level. Furthermore, the result of this study clearly showed that monthly rainfall in the majority of the stations experienced high spatial variability compared to the seasonal and annual timescales. In contrast, trend analyses of different hydrological variables showed that streamflow in most stations has changed significantly. A decreasing trend in dry, short, main rainy seasons and annual totals is dominant in six out of the nine stations, located in the semi-arid areas of the basin. The significant decreasing pattern of streamflow is observed in the eastern and northern part of the basin where land degradation is very high. Only one station, located at the basin outlet, exhibited a significant positive trend during both the dry and short rainy seasons. The different trend in this station is likely due to the construction of Tekeze hydropower dam in 2009. The remaining two out of the nine stations stayed constant in all seasons. Findings from both MK and Pettitt tests are consistent in all seasons and stations, but the timing of change points is different for the most station. This could imply that the level of human interference and physiographic characteristics is varying from sub-catchment to sub-catchment and hence high variability in runoff generation response to catchment characteristics.

Surprisingly, our results showed that there is no linkages/pattern between the trends in rainfall and streamflow in the basin. This suggests that the change in streamflow is influenced by factors other than rainfall. A weak relationship between rainfall and streamflow leads to the conclusion that the significant trends in streamflow could be due to significant changes over time of catchment characteristics, including land use/cover change, catchment management interventions and water abstractions in the upstream. This was also supported by the existing some literature as discussed in section 4.3.4.

The findings from this study are useful as a pre-requisite for studying the effects of catchment management dynamics on the hydrological variabilities. Statistical trend analyses investigate only the trend of historical data without being able to identify the causes of those trends. Therefore, further investigations are needed to verify and quantify the hydrological changes shown in statistical tests by identifying the physical mechanisms behind those changes.

Chapter 5

QUANTIFYING LONGITUDINAL LAND USE/COVER CHANGE IN GEBA CATCHMENT [3]

Spatiotemporal variability in Land Use/Cover patterns can exert a powerful influence on the hydrological processes of semi-arid areas like the Upper Tekeze basin. Accurate information about the land use/land cover change is a prerequisite for improved land and water management. The human-induced landscape transformations in the Geba catchment, one of the main tributaries of the Upper Tekeze basin, was investigated for the past four decades (1972-2014). Separate LULC maps for the years 1972, 1989, 2001, and 2014 were developed based on satellite images, Geographic Information System (GIS) and ground information. Change detection analysis based on the transitional probability matrix was applied to identify systematic transitions among the LULC categories.

The results show that more than 72% of the landscape has changed its category during the past 43 years. LULC in the basin experienced significant shifts from one category to other categories by 61%, 47%, and 45%, in 1972-1989, 1989-2001, and 2001-2014, respectively. Although both net and swap (simultaneous gain and loss of a given LULC during a certain period) change occurred, the latter is more dominant. Natural vegetation cover, including forests, reduced drastically with the rapid expansion of crops, grazing areas and bare lands during the first two decades. However, vegetation started to recover since the 1990s, when some of the agricultural and bare lands have turned into vegetated areas. Forest land showed a continuous decreasing pattern, however, it has increased by 28% in the last period (2001-2014). In contrast, plantation trees have increased by 254% in the last three decades. The increase in vegetation cover is a result of intensive watershed management programs during the last two decades. The driving forces of changes were also discussed and rapid population growth and changing government policies were found to be the most important. The provided information is essential for improved understanding of relationships between hydrological processes and catchment characteristics in the basins.

---
[3] Based on: Gebremicael, T.G., Mohamed, Y.A., van der Zaag, P., Hagos, E.Y. (2018). Quantifying longitudinal land use change from land degradation to rehabilitation in the headwaters of Upper Tekeze basin, Ethiopia, *Sci. Total Environ.*, 622-623, 1581-1589.

5.1 INTRODUCTION

Understanding land and water management in a basin require detailed information on Land Use/Cover (LULC) management practices (Woldesenbet et al., 2017; Sanyal et al., 2014; Kiptala et al., 2013; Cheema & Bastiaanssen, 2010). LULC change in combination with unsustainable land management directly impacts biodiversity, ecosystem services, crop productivity, hydrology (Arirti et al., 2015; Kiptala et al., 2013; Teferi et al., 2013; King et al., 2005), and may even influence local climate if it occurs at relatively larger scales (Wang et al., 2015). For example, changes in LULC of a catchment modify the availability of water resources of a basin in complex ways (Liu et al., 2005), and in particular the partitioning of rainfall to overland flow, interflow, and deep percolation (Taniguchi, 2012). The impact of LULC on the environmental processes is not universal and depends on the local context in a particular region (Chen et al., 2016; Lu et al., 2015; Haregeweyn et al., 2014). For example, vegetation cover reduces direct runoff volume, as well as the flow peaks, whilst enhancing the infiltration rate and hence the base flow of the stream (Yair & Kossovsky, 2002). In contrast, other studies reported reduced direct runoff and lower base flow because of the increased vegetation cover, e.g., Ott & Uhlenbrook (2004), Sanyal et al. (2014). This emphasizes the need to study both natural and anthropogenic processes, and how these may influence the environmental processes of a given basin.

The literature shows that anthropogenic factors are the main drivers of changes in land management practices and variability in water availability in Ethiopia (Bewket & Sterk, 2005). The drastic increase of population (more than doubled from 1984 to 2012) and economic development of Ethiopia escalates the pressure on natural resources (CSA, 2008). The need to provide food, water and shelter for people and livestock coupled with poor land management has significantly contributed to land degradation in the highlands (Hurni et al., 2005). A substantial increase of agricultural and grazing lands during the last few decades was at the expense of forest and shrub lands (Ariti et al., 2015; Haregeweyn et al., 2014; Teferi et al., 2013). Such changes in land surface reduced land productivity and increased water scarcity in most parts of Ethiopia (Teferi et al., 2013; Wondie et al., 2010). For example, Gebremicael et al. (2013) demonstrated that change in LULC in the Upper Blue Nile basin increased surface runoff by 75% and decreased the dry season flow by 25% during 1973-2005. However, in recent decades, the country has made significant efforts to rehabilitate degraded lands through integrated watershed management (IWSM) programs (Teferi et al., 2013; Nyssen et al., 2010). The IWSM increased vegetation cover and decreased land degradation (Belay et al., 2014; Alemayehu et al., 2009).

Local studies in the Geba catchment of the Upper Tekeze basin revealed that no specific pattern of LULC change is observed. Belay et al. (2014) reported increased vegetation cover whilst Alemayehu et al. (2009) and Bizuneh (2013) showed opposite results. However, most of those studies were conducted in a very small watershed (145-275 km²) which cannot represent the entire catchment. A limitation reported by these studies is that only net change was considered (not swap change), which could underestimate the amount of the LULC change. Pontius et al. (2004) and Teferi et al. (2013) showed that swap change (simultaneous gains and losses among LULC categories) is more important in recognizing the total change than the net change.

Therefore, the main objective of this study was to accurately assess the spatiotemporal dynamics of LULC and the associated land management changes in Geba catchment, being the source region of the Upper Tekeze basin. This basin is known for its severe land degradation and simultaneously for best experiences in environmental rehabilitation programs (Nyssen et al., 2010). Therefore, studying the patterns of LULC changes in such a contrasting environment has the potential to improve our understanding of land and water management of the basin.

5.2 STUDY AREA DESCRIPTION

The Geba catchment (5,085 km^2) is located in northern Ethiopia which extends from 38°38' to 39°48'E and 13°14' to 14°16'N (Figure 5.1). It forms the headwaters of the Upper Tekeze River basin, one of the major tributaries of the Nile River. The topography is characterized by highlands and hills in the north and north-eastern and plateaus in the central part of the catchment. The central plateaus are dissected by several rivers that flow towards the southwestern part of the basin and joins the main Tekeze River at Chemey. The altitude varies from 3,300 meter above seas level (m.a.s.l.) at Mugulat Mountains near Adigrat town to 930 m.a.s.l. at the basin outlet.

The catchment has four main sub-catchments: Siluh, Genfel, Illala and Agula. Siluh (960 km^2) sub-catchment drains the Mugulat Mountains in the northern part of Geba, with annual rainfall varying between 500 in the lower valley to more than 650 mm/year in the highland areas near E/hamus and Mugulat (Figure 5.1). Similarly, Agula (481 km^2) sub-catchment is characterized by highly dissected and rugged terrains with elevations varying from 1,750 m.a.s.l. at the confluence with the Geba, to 2,800 m.a.s.l. in the mountains near Atsbi town, with a high rainfall variability, ranging from 450 to 700 mm/year. The Genfel (730 km^2) sub-catchment is located in-between the Siluh and Agula sub-catchments, with an elevation range from 1,780 m.a.s.l. at the confluence with Siluh, 7 km upstream of Geba2 gauging station in the Geba river, to more than 2,800 m.a.s.l. in the Atsbi highlands. The fourth sub-catchment is Illala (340 km^2), which drains an area of 340 km^2 and joins the main Geba River at 2 km downstream of the Geba2 gauging station. Unlike in the earlier sub-catchments, this sub-catchment is dominated by flat areas where agriculture and settlements are the major land cover. The largest city (Mekelle) in northern Ethiopia is found within this sub-catchment. The rainfall over this catchment is very erratic in distribution and magnitude with an annual average below 550 mm/year. Further down the Geba river is joined by smaller tributaries.

The Geba catchment is characterized by a semi-arid climate in which the majority of the rainfall occurs from June to September after a long dry season. More than 70% of the total rainfall is falling in July and August only with high storm intensities (Chapter 4). Rainfall over the catchment is highly variable mainly associated with the seasonal migration of the intertropical convergence zone (ITCZ) and the complex topography (Nyssen et al., 2005). The mean annual precipitation ranges from below 440 mm/yr in the Eastern part to 800 mm/yr in the Northern and the Western part of the basin.

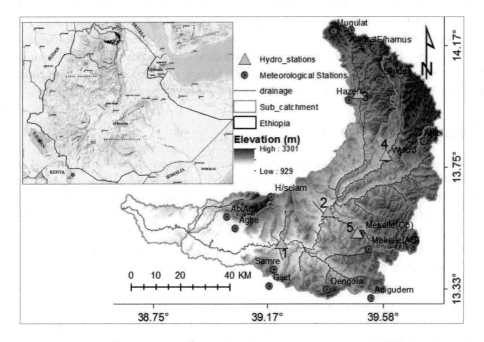

Figure 5.1: Location map of the Geba catchment with its elevation (m.a.s.l) in the upper Tekeze river basin (1=Geba1, 2=Geba2, 3=Siluh, 4=Genfel and 5= Illala)

Land use in the basin is dominated by rainfed agriculture (40%) followed by shrubs (27%), bare land (20%), grassland (9%), forest (1%) and residential areas (3%). Common rainfed agriculture in the basin includes Teff, wheat, barley, maize, sorghum and pulses. However, irrigated agriculture at the household level has also increased significantly in the recent years (Nyssen et al., 2010; Alemayehu et al., 2009). Bare land and shrubs are mainly dominant in the semi-arid eastern lowlands of the basin while most of the cultivable lands and very small forest coverage occur in the highlands but dry areas of the basin. Agricultural and bare lands have been expanded at the expense of all other land uses in the basin.

The majority of the geological formations are Enticho Sandstone, Edag Arbai Tillites, Adigrat Sandstone, Antalo Super sequence and Metamorphic (basement) rocks (Gebreyohannes et al., 2013; Virgo & Munro, 1977). The dominant soil types of the catchment are clay loam, sandy clay loam, clay, loam and sandy loam (Abraha, 2014; Zenebe et al., 2013). The occurrence of soil textures across the whole catchment is deeply weathered soil in the uppermost plateaus, rocky and shallow soils in the vertical scarps, coarse and stony soils in the steep slopes, finer textured soils in the undulating pediments and most deep alluvial soils are found in the alluvial terraces and lower parts of the alluvial deposits (Abraha, 2014; Gebreyohannes et al., 2013). Most of the soils in the basin are limited in depth due to contagious hard rocks and cemented layers and hence not suitable for agricultural production despite farmers having cultivated them for a long time because of shortage of arable land in the basin (Virgo & Munro, 1977).

This catchment is known for its severe land degradation resulted from the high population and socio-economic developments. However, the recent IWSM programs have led to significant improvements of natural resources. Various forms of intervention programs including; physical Soil and Water Conservation (SWC), biological SWC, water harvesting programs and conservation agriculture have been implemented in the basin (Belay et al., 2014; Nyssen et al., 2010; Alemayehu et al., 2009; Munro et al., 2008). These programs have been implemented by mobilizing the community for free-labour days and food for work through different programs including; Managing Environmental Resources to Enable Transition to more sustainable livelihoods (MERET), Productive Safety Net Program (PSNP), and the National Sustainable Land Management Project (SLMP), (Haregeweyne et al., 2015; Munro et al., 2008).

5.3 DATA AND METHODS

5.3.1 Data acquisition

Satellite images for the years 1972, 1989, 2001, and 2014 were acquired to analyse the spatiotemporal patterns of the LULC in the Geba catchment of the Upper Tekeze basin. Landsat images for the selected years were obtained from the US Geological Survey (USGS) Centre for Earth Resources Observation and Science (EROS) found in http://glovis.usgs.gov/. A brief description of the collected satellite image including, satellite and sensors used, resolution, acquisition date, swath/row of each images for each period is summarized in Table 5.1. The years for analysis were selected based on key signs of LULC change, e.g. land degradation, land policy changes, the IWSM interventions, and finally, the availability of satellite image. Images from the same period (October-December), i.e. immediately after the rainy season, were selected in order to minimize the seasonal effect on the classification results (Wondie *et al.*, 2011). Ancillary data, including field data (ground truth points (GTPs), interviews and observations), topographic maps, areal photographs, and secondary literature were collected to derive reference data and topographic variables.

The GTPs used for the classifications and accuracy assessment was collected during a field survey from 12/12/2015 to 29/06/2016, as well as from the ancillary data. The GTPs for the 2014 image were collected using Geographic Positioning System (GPS) whilst for 2001, 1989 and 1972, these were derived from available areal photographs (1994, 1986 and 1965), and existing local and national LULC maps (Belay *et al.*, 2014; HTSL, 1979). In-depth interviews were conducted with 120 local elderly farmers to collect LULC information on 2001, 1989 and 1972. A digital elevation model (DEM) from Shuttle Radar Topographic Mission (SRTM) with

30 m resolution and topographic maps of scale 1:50,000 from the Ethiopian Mapping Agency (EMA) were also collected for geo-referencing and deriving topographic variables.

To determine the impact of socio-economic development and IWSM interventions on LULC change, population data and historical information of IWSM were collected from the Central Statistical Agency of Ethiopia (CSA), the Bureau of Agriculture and Rural Development (BoRD) of the region and from a field survey through interviews of local elders.

Table 5.1: Specifications of satellite images used for this study

Satellite	Sensor	Path/row	Resolution (m)	Acquisition date
Landsat 1-3	MSS Multispectral	181/50/51	60	02-11-1972
Landsat 4-5	TM Multispectral	168/51 169/50/51	30	21-11-1989
Landsat 7	ETM+ Multispectral	168/51 169/50/51	30	16-11-2001
Landsat 8	Operational Land Imager (OLI)	168/51 169/50/51	30 *	04-12-2014 11-12-2014

5.3.2 Image pre-processing

Image pre-processing (geometric and atmospheric corrections and topographic and temporal normalizations) was done for all LULC maps. This was vital to establish a direct association between the acquired data and physical occurrences of the study area (Hassan et al., 2016). All satellite images and digitized ancillary paper maps were georeferenced to a common coordinate system using the topographic map at 1:50,000 scale and Universal Transverse Mercator (UTM) map projection (Zone 37) and WGS 84 datum in ArcGIS 10.1 and Erdas Imagine 2014. Even if the collected satellite data were georeferenced from the sources, it was necessary to geo-rectify them against the base map using 60 control points in order to avoid some misalignments in the study area

The study area is characterized by rugged topography and topographic and atmospheric corrections of the images were needed to retrieve the actual surface reflectance. The reflectance in rugged terrains is strongly influenced by solar illumination and viewing geometry. Moreover, the terrain illumination correction is important in the time series imagery analysis as its effect may vary with different dates (Kopačková et al., 2012). Such effects were minimized using Topographic and Atmospheric Correction for Airborne Imagery (ATCOR-4) package (Richter & Schlapfer, 2012). This method is a well-established process for atmospheric and topographic corrections using DEM (Kopačková et al., 2012). High accuracy of atmospheric compensation and topographic correction is achieved by employing the correlated k algorithm in the ATCOR-4 software. Finally, normalization of pixel brightness value variations among the images taken in the different period was done by applying image regression (Jensen, 1996).

5.3.3 Image classifications, accuracy assessments and post classifications

A hierarchical LULC classification was derived based on descriptions of US Geological Survey LULC approach (Anderson et al., 1976.), the author's prior experiences to the study area and local studies (Belay et al., 2014; Alemayehu et al., 2009). Descriptions of identified major LULC classes are given in Table 5.2.

First, unsupervised classification was performed to identify clusters by their spectral similarities. An Iterative Self-Organizing Data Analysis Technique (ISODATA) clustering algorithm which is the most common for clustering of pixels was selected to identify the spectral classes (Congcong et al., 2014). This algorithm is the most commonly used technique for cluster grouping of pixels during unsupervised classifications. It compared each candidate pixel to the cluster mean by lowest spectral distance method and a pixel with closer distance to the candidate pixel is assigned to the cluster. Final classifications were done using ISODATA algorithm based on a 96% convergence threshold.

Table 5.2: Descriptions of identified LULC of the study area

LULC class	Description	Code
Grassland	A land dominated by natural grass, small herbs and grazing lands	GRAS
Agriculture	Areas covered with annual and perennial crops covering more than 70% of the land.	AGRI
Bushes and shrubs	Areas covered by low woody vegetation of less than 3m in height, multiple stems, vertical growing of bushes and shrubs with canopy cover between 5-50%.	BUSH
Wooded bushes	An intermediate b/n forest and bushes with >3m in height and less dense vegetation	WOOD
Settlements	Built up areas including towns, cities and small villages	URBA
Water body	Any types of surface water	WATE
Bare land	Little or no vegetation cover, exposed rocks and degraded land	BARE
Forest land	High and dense natural forest around churches and reserved areas	FORE
Plantation forest	Forests planted by man	PLAN

The LULC maps were further refined with expert judgment and field observations using supervised classifications. Information on 3,326 GTPs was collected from different field observations. Out of this, 1,186 points were used for classifications while the remaining 2,140 points were used for accuracy assessment. These GTPs were sampled by a stratified random design which is the most common technique to attain high overall accuracy (Congalton, 2001). The sample size of GTPs required for each LULC was assigned based on the dominant classes and inherent variability within the group. Training sites corresponding to each classification type were taken among the collected samples. Signatures for each LULC type were developed and their separability was evaluated based on Euclidean distance. Accordingly, it was found that bare land was difficult to distinguish from urban areas and, to avoid the confusion, urban areas were masked out from the image and classified independently. In addition, there was some confusion between bare and arable lands and plantation and forest classes in the preliminary classification, but these uncertainties were minimized by adding more training samples for each class. Finally, the Maximum Likelihood Algorithm which has superior performance compared to other approaches was used for classification (Al-Ahmadi & Hames, 2009).

To determine the degree of acceptability of the classification processes, accuracy assessment is an important task in remote sensing image analysis (Congalton, 2001). In this study, a holistic accuracy assessment approach using GTPs, validation of land use type with local datasets and visual inspection was adopted. First, visual inspections were performed considering the first author's knowledge of the study area. This evaluation method is a necessary first step in checking the quality of classified maps (Foody, 2002).

Next, the final LULC maps were evaluated using independent GTPs. The accuracy of classified maps was evaluated by computing an error matrix which compares the classification outputs with the GTPs. Quantitative accuracy assessment based on an error matrix is an effective way of evaluating the quality derived maps (Foody, 2002). By comparing the maps and collected 2,140 GTPs, characteristic coefficients (User's and Producer's accuracy, overall accuracy and Kappa coefficients) were obtained from the confusion matrix. The User's and Producer's accuracy were used to estimating the accuracy of each LULC types whilst the overall accuracy was applied to estimate the overall mean of user and producer accuracy. The Kappa coefficient expresses the agreement between two datasets corrected for the expected agreement (Van Vliet et al., 2011). Further validations of these LULC maps were done using local datasets and field observations. Information from local LULC change pattern studies (Belay et al., 2014; Alemayehu et al., 2009) was used to cross-check the results. Moreover, field visits to 80 randomly selected sites were done to verify whether the classified LULC types were the same at ground level. The maps were further verified from the pre-determined sites through observations and consulting of 42 local elders.

The post-classification change detections between two independent LULC maps were performed in ArcGIS 10.1 (Jensen, 2005). The developed LULC maps of 1972, 1989, 2001 and 2014 were overlaid two at a time and converted areas from one particular class to any other classes were computed. Transition matrices were computed to analyse the LULC change between two independent maps. To analyse the rate of change, location and nature of the changes, the gains, losses, net change (Ng), total change (Tg), persistence, gain to persistence ratio (Gp) and loss to persistence ratio (Lp) and swap change (Sw) were computed. Detailed descriptions of such calculations can be found in Yuan et al. (2016). These indicators can give robust information on the resistance and vulnerability of a given LULC type (Braimoh, 2006). Patterns of degradation and recovering were carefully examined and compared among the study periods.

To understand the possible drivers of changes, information on socio-economic and environmental factors, including population growth, drought and famine events, government policy changes and environmental rehabilitation programs were analysed using qualitative and descriptive statistics. The relationships between these potential driving forces and land degradation or rehabilitation programs were examined by describing the state of change in these factors and the reactions of the LULC type over time. To find out the contribution of watershed management interventions, analysis of LULC changes before and after IWSM was done in representative small watersheds. The selected watersheds were taken from different locations where intensive IWSM has been implemented since the early 1990s.

5.4 RESULT AND DISCUSSION

5.4.1 Classification and change detection

The results of the LULC classification for the years 1972, 1989, 2001 and 2014 are shown in Figure 5.2. The accuracy of the developed maps was evaluated using statistical indices described in previous sections. The confusion matrix report for the classification maps of 2014 is presented in Table 5.3. The overall accuracy and Kappa coefficient of the 2014 map were 84.3% and 81.1%, respectively. Similarly, overall accuracy (and Kappa coefficient) of 86.5% (84.3%), 86.2% (82.6%) and 84.3% (80.4%) were obtained for the maps of 2001, 1989 and 1972, respectively. Confusion matrices for the remaining maps (1972, 2001 and 2001) are also presented in Appendix C (Table C-1). The User's and Producer's accuracies of all maps exhibited a relatively good accuracy of more than 80% for the majority classes. A lower individual accuracy was obtained for plantation (72%) and agriculture (76.4%) which may be associated with the confusion of reflectance with forest and bare land, respectively. Kappa coefficient greater than 80% represents a strong agreement between classified classes and the reference data (Ariti et al., 2015; Kiptala et al., 2013). Obtained statistical indices demonstrated that the produced maps are sufficiently accurate for further LULC change detections.

The produced maps and estimated area of each LULC category are presented in Figure 5.2 and Figure 5.3a, respectively. The results showed that bushlands, agriculture, woodland and bareland were the dominant LULC types at the beginning of the study period. However, as indicated in Figure 5.3b, wood, forest, grass and bush lands significantly declined while bareland and agriculture increased during the 1972-1989 period. This indicated that there was a substantial reduction of vegetation cover in this period to obtain more crop and grazing lands. Furthermore, in 1989, the plantation and water body emerged as new land cover classes.

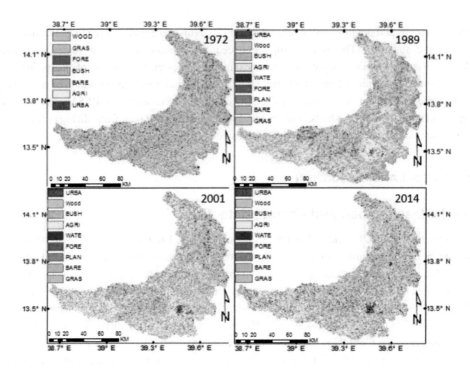

Figure 5.2: Land use and land cover maps of the Geba catchment in 1972, 1989, 2001 and 2014.

Table 5.3: Confusion matrix for 2014 classification map

Classificat ion result	Reference (field) data									row totals	User's acc. (%)
	WATE	AGRI	BARE	FORE	PLAN	WOOD	BUSH	GRAS	URBA		
WATE	54	0	0	0	0	0	0	0	0	54	100.0
AGRI	0	414	44	3	7	5	49	18	2	542	76.4
BARE	2	35	235	0	5	6	6	5	0	294	79.9
FORE	0	2	0	147	39	0	2	0	0	190	77.4
PLAN	0	0	0	15	155	2	2	0	0	174	89.1
WOOD	0	0	2	15	5	192	5	0	0	219	87.7
BUSH	0	18	6	2	2	6	293	3	2	332	88.3
GRAS	0	12	0	2	2	0	2	239	0	257	93.0
URBAN	0	0	0	0	0	0	1	2	75	78	96.2
Column totals	56	481	287	184	215	211	360	267	79	2140	
Producer's Acc. (%)	96.4	86.1	81.9	79.9	72.1	91.0	81.4	89.5	94.9		
Over all accuracy = 84.3%, Kappa coefficient = 81.1%											

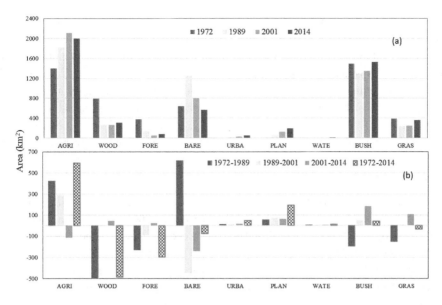

Figure 5.3: Area coverage for each LULC (a) Absolute area of each LULC for the four periods, (b) Change in areas between each time step (note that the last bar of each LULC type is the total change for the entire study period, 1972-2014)

During the period 1989-2001, the largest decreasing rate was observed in forest cover (-63%) which reduced from 145 km^2 in 1989 to 55 km^2 in 2001 and bare land (-36%) reduced from 1250 km^2 to 800 km^2 with a slight decrease (from 262 to 257 km^2) in woodland whereas the other categories showed a relative growth. The observed decrease in bareland and a slight increase in bushland during this period reflects the government policy of the region towards recovering the environment after the devastating drought of 1984/1985. Despite the efforts made to rehabilitate the environment, deforestation of natural forests in highland areas of the basin continued at a significant rate during this period (Figure 5.2 and Figure 5.3a). Although the water body is hard to appraise from Figure 5.2 and Figure 5.3, it has significantly increased from less than 0.25 km^2 in 1989 to 7 km^2 in 2001. The reason is most likely due to the attention given to small and medium scale water harvesting programs in the 1990s (Haregeweyn et al., 2006). Similarly, urban settlements also significantly increased (from 12 to 31 km^2) during this period, which may be associated with population growth and industrial development around towns.

In the period from 2001 to 2014, all LULC categories except agriculture and bare land showed a relative increase ranging from 4% to more than 90% (Figure 5.3b). A slight decrease (-5%) in agriculture during this period is due to the intensification of urban areas and homestead plantations at the expense of cultivable land. Although at a slower rate compared to the previous period (1989-2001), bare land continued to decrease (from 800 to 561 km^2) during this period. One possible reason is the impact of environmental rehabilitation programs which have been intensively implemented in the last two decades (Nyssen et al., 2010; Belay et al., 2014). The positive change (14%) in bush land during 2001-2014 was more pronounced compared to the period 1989-2001 which was only 4%. Significant increases of wood (18%) and forest lands (28%) during this period indicates the emergence of new bush and shrublands while existing ones were transformed into wood and forest land. Urban and the water body areas that have started to increase in the preceding period continued the increasing trend from 31 to 48 km^2 (57 %) and from 6.4 to 12 km^2 (92%), respectively. Overall LULC in the catchment has considerably changed in the last 43 years. Plantation, urban and water body showed a continuous increasing trend in all periods but the remaining LULC categories exhibited different change patterns in different periods.

Figure 5.4 explicitly shows how the area of each LULC has fluctuated for the entire study period. Agricultural land increased in the 1972-1989 (30%) and 1989-2001 (16%) but started to decline in 2001-2014 by -5%. In contrast, bush and grasslands decreased by 13% and 39%, respectively in 1972-1989 then slightly increased in the 1989-2001 and 2001-2014 periods. Over the whole period (1972-2014), agriculture, urban areas, bush and shrubs, plantations and surface water showed a relative increment whilst the remaining LULC types exhibited a decreasing trend (Figure 5. 3b and Figure 5.4). The largest and smallest percentage of positive changes were observed in urban area (4874%) and bushes (2.8%), which increased from 1 to 48 km^2 and from negligible to 12 km^2 respectively while forest and Grass land showed the largest (-79%) and smallest (-8.4%) decreasing percentage for the entire period. A remarkable result is that forest land showed a continuous decreasing trend until 2001, but then increased from 56 km2 in 2001 to 78 km^2 in 2014.

The transitional probability matrix of LULC in Table 5.4 shows the sources and direction of the different changes in the period between 1972 to 1989 and 1972 to 2014. Transitional probability matrix for the period between 1989 to 2001 and 2001 to 2014 are also presented in Appendix C (Table C-2). The matrix revealed that it has been a considerable and dynamic LULC change in the study area for the last 45 years. Most of the LULC categories experienced both positive (increase) and negative (decrease) changes. The diagonal and off-diagonal elements show the area of LULC that have persisted and changed for the given period, respectively. The landscape in the Geba catchment has shown a shift from one category to other different categories by 61%, 46.9%, 44.5% and 72.2% during the period 1972-1989, 1989-2001, 2001-2014 and for the entire study period 1972-2014, respectively. On the other hand, 39%, 53.1%, 55.5% and 27.8% of the total landscape remains unchanged for the same period which clearly indicates the persistence of LULC dominates in the second and third period while the largest proportion of change in landscape was experienced during 1972-1989 and 1972-2014 periods.

Both swap and net changes which are very important in recognizing the total changes within the landscape occurred in all periods. About 39%, 36%, 37.3% and 54.4% of the total change were found due to swap changes in 1972-1989, 1989-2001, 2001-2014 and 1972-2014 time period, respectively. The changes attributed to swap change were higher than the net changes in all stages. Only 22.2%, 11.3%, 7.0% and 17.7% of the total changes for the period of 1972-1989, 1989-2001, 2001-2014 and 1972-2014, respectively, were observed from the net changes. This explicitly shows that there were a coinciding gain and losses among the LULC categories in all periods.

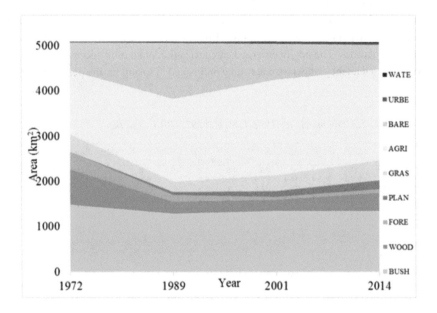

Figure 5.4: Evolution of LULC change in the Geba catchment, 1972-2014

Analysis of LULC patterns based on the net change, especially in the semi-arid areas underestimates the total changes as it fails to account the swap changes (Pontius et al., 2004). For example, the total change (72.2%) of the landscape for the whole study period (1972-2014) would have been underestimated by the net change (17.7%) which was later accounted in the swap change (54.4%). This implies the swap change accounted for the LULC categories lost in one site and an equivalent dimension is added to different sites.

During the whole period (1972-2014), forest, agricultural and bushlands have experienced a relatively higher persistence of 38.05%, 34.3%, and 32.2%, respectively, while the grassland (15.5%) and bare land (15.2%) categories showed a lower percentage of persistence for the same period. A relative higher persistence in agriculture and bushland is mainly due to their higher proportion in 1972 (Table 5.4 and Fig.5.3a). The higher percentage of persistence in the forest could be due to the preserved Dessia forest in Eastern part of the basin and church forests across the basin. The gain-to-persistence (Gp) and loss-to-persistence (Lg) ratios shown in Table 5.4 indicates the likelihood of LULC categories to gain or loss comparing to its initial, respectively. Accordingly, most LULC categories showed greater than 1 of Gp and Lp during the study period (1972-2014) which implies these categories have a tendency of gaining or losing rather than persisting. For example, a greater Gp ratio is observed in urban and barelands comparing to other categories which suggest that they have a higher tendency of gaining rather than persisting. Similarly, a higher loss-to- persistence ratio was found in forest, wood and bare lands. This clearly demonstrates a higher tendency of transition to other LULC categories is dominated in these classes. On the other hand, observed lower Lp in urban and agricultural lands in comparison to other categories shows unchanged extent and minimum losses. This is true for the observed intensified agricultural cultivations with drastic increase of the population in the last four decades. The newly emerged categories (i.e. plantation and water) in 1989 maps has also shown a tendency of persistence and gain instead of losing for the period of 1989-2014.

Table 5.4: *LULC transition matrix for 1972-1989 and 1972-2014, area (km²)*

		AGRI	BUSH	BARE	GRAS	WOOD	FORE	URBA	PLAN	WATE	totals	loss	Sw	Lp
							To final state in 1989							
	AGRI	880.7	107.8	335.2	53.2	0.2	0.8	8.1	16.4	0.1	1402.2	521.5	1043.1	0.6
	BUSH	323.5	663.8	326.4	58.4	77.5	29.8	0.8	9.2	0.0	1489.2	825.5	1263.9	1.2
	BARE	244.9	143.6	198.5	20.2	24.8	0.1	0.0	3.5	0.1	635.6	437.1	874.2	2.2
Initial state from 1972	GRAS	67.0	43.4	158.4	87.3	20.5	0.5	2.5	11.5	0.0	391.0	303.7	304.7	3.5
	WOOD	227.9	256.1	151.8	16.1	82.4	41.9	0.0	14.4	0.0	790.7	708.3	358.9	8.6
	FORE	80.4	81.1	79.6	4.4	56.6	72.3	0.0	0.6	0.0	374.9	302.6	146.2	4.2
	URBE	0.0	0.0	0.0	0.0	0.0	0.0	1.0	0.0	0.0	1.0	0.0	0.0	0.0
	PLAN	0.0	0.0	0.0	0.0	0.0	0.0	0.0	0.0	0.0	0.0	0.0	0.0	
	WATE	0.0	0.0	0.0	0.0	0.0	0.0	0.0	0.0	0.0	0.0	0.0	0.0	
	totals	1824.3	1295.7	1249.8	239.6	261.9	145.4	12.3	55.5	0.2	5084.7	3098.7	1995.5	1.6
	gain	943.6	631.9	1051.3	152.4	179.5	73.1	11.4	55.5	0.2	3098.7			
	Ng	422.1	193.5	614.2	151.4	528.9	229.5	11.4	55.5	0.2	1103.3			
	Tg	1465.1	1457.4	1488.4	456.1	887.8	375.7	11.4	55.5	0.2	3098.7			
	Gp	1.1	1.0	5.3	1.7	2.2	2.0	11.8	NA	NA	1.6			
							To final state in 2014							
	AGRI	662.5	324.6	177.4	109.7	0.5	0.3	38.5	78.4	6.9	1398.8	736.3	1472.6	1.1
	BUSH	601.2	439.4	90.1	112.6	177.9	18.1	2.9	46.6	2.6	1491.5	1052.0	1804.3	2.4
Initial state from 1972	BARE	319.8	174.9	87.4	31.7	0.9	6.1	0.0	14.2	1.6	636.6	549.2	946.2	6.3
	GRAS	87.0	75.0	45.9	83.5	55.6	0.9	0.0	43.4	0.7	392.0	308.5	617.1	3.7
	WOOD	240.5	228.6	106.6	69.4	110.1	23.4	0.0	12.8	0.4	791.8	681.7	599.7	6.2
	FORE	84.4	99.1	53.2	40.3	65.0	30.0	0.0	0.5	0.0	372.4	342.4	97.7	11.4
	URBE	0.0	0.0	0.0	0.0	0.0	0.0	1.0	0.0	0.0	1.0	0.0	0.0	0.0
	PLAN	0.0	0.0	0.0	0.0	0.0	0.0	0.0	0.0	0.0	0.0	0.0	0.0	NA
	WATE	0.0	0.0	0.0	0.0	0.0	0.0	0.0	0.0	0.0	0.0	0.0	0.0	NA
	totals	1995.4	1341.6	560.4	447.2	409.9	78.9	42.4	196.0	12.2	5084.1	3670.2	2768.4	2.6
	gain	1332.9	902.2	473.1	363.7	299.8	48.9	41.4	196.0	12.2	3670.2			
	Ng	596.6	149.9	76.2	55.2	381.9	293.6	41.4	196.0	12.2	901.4			
	Tg	2069.2	1954.2	1022.3	672.3	981.5	391.3	41.4	55.5	0.2	3670.2			
	Gp	2.0	2.1	5.4	4.4	2.7	1.6	41.0	NA	NA	2.6			

The LULC transitions are complemented by gains and losses among each LULC categories in all periods. All categories except urban areas, plantation forests and water body consist of both net and swap changes for the whole study period. On the contrary, only net change occurred in the remaining classes. For example, most of the total changes in bushland (56.4%), agriculture (53.2%) and bareland (27.9%) during the 1972-2014 periods were attributed to swap changes by 49.2%, 40.1%, and 25.8%, respectively. This indicates only 16.3%, 4.15 and 2.1% of the total changes for each class is occurred by net changes. During the 1972-1989 periods, swap changes were dominated in the bush land (40.6%), agriculture (33.7%) and bare land (28.8%)

comparing to the net change of 6.2%, 13.6% and 20.1% for each category. Similarly, both swap and net types of change occurred in the 1989-2001 period. However, the contribution of swap changes was found to be the most dominant in agriculture, bushland, bareland and grasslands whilst net change was the main driver of the total change in the remaining categories. Changes in the urban areas and water body in this period continued to have almost pure net change.

In the period 2001-2014, all categories with the exception of the urban area and water body consisted of higher swap changes. In summary, higher swapping than net changes was observed in most categories during all periods. This explicitly shows that these LULC were the most dynamic LULC categories in the study area. Interestingly, the multiple-sided swap change among most LULC categories underlined the damage and simultaneously recovering of landscapes in the sources region of Tekeze basin.

The most important finding of this study is that there were periodical fluctuations on the rate of transitions among the LULC categories throughout the study period. The rapid expansion of agricultural and barelands during the first study period (1972-1989) was at the expense of natural vegetation. Conversion of bushes and shrubs, wood land and forest to agricultural area accounts about 34.3%, 24.4%, 8.5%, of the total converted area. However, there was also a substantial exchange of areas between bareland and agricultural lands during this period. Systematic transitions from all categories to agricultural lands occurred during this period which implies that the agricultural land has systematically gained from all LULC classes. Growing pressure from human and livestock population might have significantly contributed to rapid expansion of agricultural lands at the expense of vegetation cover. The agricultural area has continued to systematically gain from all LULC classes during the second (1989-2001) and third period (2001-2014). However, the rate of clearing forest and woodlands for agriculture expansion has slowed down and the majority of the gaining was from bare, grass and to some extent bush and shrub lands. On the other hand, vegetation cover has started to recover since the early 1990s where some of the agricultural, bare and grass lands have turned into plantations and bushes and shrubs. These plantations were established through government and international aid programs following the increased demand in construction and fuel woods (Tekle & Hedlund, 2000; Zeleke & Hurni, 2001). For example, 65% of the total plantations in 1989 to 2001 were obtained by abandoning arable, grass and bare lands. However, plantations of Eucalyptus trees on agricultural lands has slowed down after 2001 and more focus was given to enrichment plantations in non-agricultural lands. Possible reasons for declining pattern of plantations in agricultural land are its negative effect on the water table and soil fertility (Belay et al., 2014; Leite et al., 2010).

Summarizing over the whole study period (1972-2014), agricultural land has systematically gained and loss to all LULC categories (Table 5.4). Bush and shrubs, wood and forest lands consisted of more than 70% of the total agricultural gains that have systematically transformed from 1972 to 2014. At the same time, agricultural areas lost due to swap and net change during this period have systematically transformed into all categories with the exception of forest and wood land. The prominent transitions from agriculture were to bush and shrubs (44%), bareland

(24%), grass (14.8%), urban areas (10.8%) and plantations 5.2%). The reverse transitions, i.e. a slight decreasing of agricultural areas and increasing of vegetation cover during the 2001-2014 period, indicates the recent efforts made to environmental rehabilitation programs. More than 92% of the total urban areas obtained from arable land which indicates expansion of urban settlements is also one of the possible reasons for the declining of agricultural lands during the study period.

Lack of consistent trend in most LULC categories during the different period evidently shows how the land use/cover change over the basin was dynamic. In summary, this study confirms that expansion of agricultural and grazing areas has been the primary causes of losses in natural vegetation in the semi-arid areas of Tekeze basin. Increasing demands in construction and fair wood following population and economic growth in the last three decades could have also contributed to the loss of natural vegetation. This finding is in agreement with some local studies (e.g. Abraha, 2014; Belay et al., 2014; Bizuneh, 2013; Alemayehu et al., 2009;) who reported that expansion of agriculture was the main driving force for declining of natural vegetation.

5.4.2 Possible drivers of the land use/cover changes

Identifying the major drivers is very helpful in understanding the reasons behind the change patterns of LULC in any basin. Natural environment, demographic, socio-economic and infrastructure developments are the most important factors which can significantly influence the LULC dynamics (Wondie et al., 2011; Tsegaye et al., 2010; Wang et al., 2010; Lambin et al., 2003). The response of most LULC transitions to population growth, government policy changes, major historical (drought) events and economic growth within the study area were carefully analysed and discussed in this section. These factors are found to be the most driving forces of LULC change in the Ethiopian context (Belay et al., 2014; Teferi et al., 2013; Hurni et al., 2005; Tekle & Hedlund, 2000).

The pressure of population growth on the LULC has been considered the most important change driving factor in the semi-arid region of Ethiopia (Haregeweyn et al., 2014). According to the 1994 (CSA, 1994) and 2007 censuses (CSA, 2008) and projections, the total population within the Geba catchment nearly tripled from 335,500 in 1972 to 934,290 in 2014, and population density increased from 65 to 185 persons per square kilometre. The rapid rise in population over the last 43 years shows human activities were the prominent motives behind the observed land degradation due to the expansion of farmlands, settlements and uncontrolled utilization of natural vegetation for firewood.

The increase in population within the basin was paralleled by increasing in active farmlands and decreasing of vegetation covers. Correlation analysis between the pattern of population and expansion of agricultural areas showed a higher positive correlation (r = 0.76) suggest that rapid growth of cultivable land is subjected to the population growth. Population growth between 1972 and 2014 has also shown a negative correlation with bushes and shrubs (r = -0.48), forest land (r = -0.63) and woodland (r = -0.64). This clearly showed that natural vegetation cover in the basin has significantly decreased because of rising in population. Contributions from the recent environmental rehabilitation programs could be one of the main reasons behind a weaker negative correlation between population growth and bush and shrublands (Belay et al., 2014; Descheemaeker et al., 2006a; Negusse et al., 2013). Many studies elsewhere in Ethiopia (e g. Bewket, 2002; Zeleke & Hurni, 2001; Tekle & Hedlund, 2000; Hurni, 1993) reported that population growth is the major driving force of LULC change. This finding is also supported by some local studies within the basin (Belay et al., 2014; Alemayehu et al., 2009) who showed that land degradation was caused by increasing of human activities.

Apart from population growth, the recurrent drought conditions in the region (e.g. in 1972/1974, 1984/1985 and 2002) was also perceived as the possible drivers of land degradation in the basin (Belay et al., 2014; Hurni et al., 2005; Tsegaye et al., 2010). Following those drought periods, farmers in the study area started to consider alternative sources of income to cope up with the associated adverse effects. Expansion of agricultural lands by clearing vegetation cover in steep slopes, expansion of irrigation, selling of charcoal and firewood were amongst the coping strategies of drought caused by rainfall variability. This series related impacts influenced the overall land management of the region if not the whole country. Similar studies (Alemayehu et al., 2009; Biazin & Sterk, 2013; Bewket, 2002) reported that a shift in coping mechanisms of drought significantly affected the LULC dynamics. Government's strategy and policy on land tenure and economic development might have also significantly contributed to the observed LULC changes. Expansion of agricultural lands related to a shift in government's policy has been one of the main drivers of these changes. Agricultural led industrialization economic development policy has been adopted since 1991 by the current government (Zenebe et al., 2013). The observed prompt expansion of agricultural and settlement areas after the early 1990s could be attributed to this policy and strategy changes of the government. The policy on environmental rehabilitation, and the associated implementation programs has most plausibly been the key driver behind the observed increasing trend of vegetation cover since 1989 (Nyssen *et al.*, 2010). Similar studies elsewhere in the world (e.g., Wang et al., 2010; Liu et al., 2008; Lambin et al., 2003; Bürgi & Turner, 2002) also underlined that government policy on land management results in rapid modification of landscapes and the overall ecosystems in a basin.

5.4.3 Effects of watershed management interventions on the LULC patterns

Implementation of IWSM interventions in the last two decades has significantly contributed to the recovery of vegetation cover. To establish the contribution of these interventions, LULC change before and after IWSM were analysed in three small watersheds (Abreha Atsbiha, Negash and Birki). Results from these watersheds confirm that vegetation cover has continuously increased from 1989 to 2014 (Table 5.5 and Figure 5.5). Plantation, forest, wood and grasslands in the Abreha Atsbiha watershed has increased by 140%, 75%, 420%, and 41%, respectively (Table 5.5). Most of the bare, bushes and shrublands were transformed into these LULC classes. Agricultural areas slightly decreased for the same period to allow homestead plantations. Similarly, woods and plantations significantly increased after IWSM in the Negash watershed (Table 5.5). Bareland areas in this watershed significantly declined and converted into vegetation cover. Similar patterns were observed in the Birki watershed. In summary, biological SWC interventions, particularly exclosures and plantations on steep slopes, have significantly enhanced vegetation cover which is also supported by Alemayehu et al. (2009) and Descheemaeker et al. (2006)

Table 5.5: The rate of LULC change in three watersheds after IWSM interventions, between 1989-2001, 2001-2014 and 1989-2014 (km²)

Watersheds	Period	LULC category						
		FORE	PLAN	WOOD	GRAS	BARE	BUSH	AGRI
Abreha Atsbiha	1989/2001	0.1	0.4	0.5	1.4	-1	-0.4	-0.9
	2001/2014	0.4	1.2	1.3	-1	-1	-1.8	-0.5
	1989/2014	0.8	1.6	1.4	0.9	-1.4	-2.2	-1.4
Negash	1989/2001	-0.1	0.7	0.7	1.1	-2.2	-0.6	0.5
	2001/2014	0.1	0.4	-0.7	-0.4	-0.7	0.4	0.7
	1989/2014	-0.5	1.5	2	1.6	-4.7	0.1	0
Birki	1989/2001	0	0	3.2	2.3	-2.3	-3.1	-0.1
	2001/2014	-0.4	0.4	0.1	-0.7	0.7	-0.7	0.4
	1989/2014	0.1	0.5	2.6	1.9	-3	-2.7	0.6

The results from these watersheds is evidence that the observed recent increment in vegetation cover at the watershed level may be attributed to IWSM. However, the rate of increase over the entire Geba catchment (Figure 5.2-5.4) appeared to be small compared to the treated watersheds. This suggests that significant improvements in vegetation cover in areas with intensive SWC interventions are counter-balanced by continued land degradation in areas without such interventions. In conclusion, the collective evidence from this and previous studies confirms that the observed increase in vegetation cover at the basin scale resulted from the recent IWSM programs.

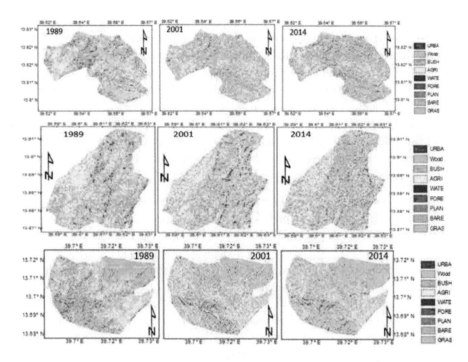

Figure 5.5: Vegetation cover before and after integrated watershed management interventions at Arbha watsbiha (top), Birki (center) and Negash (bottom)

The observed reversal in LULC dynamics in the catchment will certainly have implications for the hydrological processes both at local and basin scales. Continuously changing LULC is likely to have influenced the partitioning of rainfall into different hydrological components. Gebremicael et al. (2017) showed that variations in streamflow in the absence of significant change in rainfall patterns in the catchment could be attributed to changes in land management practices.

It is also important to highlight some of the limitations of this study as they are essential to know the negative effects on the result. Uncertainties and inconsistencies from collection devices (sensors), discrepancies in LULC class definitions and uneven distribution of reference data are some of the limitations that may have affected the result of this study.

5.5 CONCLUSION AND RECOMMENDATIONS

Detailed information on the spatiotemporal patterns of LULC is essential for understanding land and water management in a basin. Analysis of multi-temporal LULC change and possible associated drivers were done in the source region of the Upper Tekeze basin. The general trend observed in the present study is a decrease in vegetation cover and an increase in agricultural and settlement areas. The majority of LULC categories has changed substantially in the last 43 years (1972-2014). Three LULC categories (tree plantation, urban and water body) exhibited an increasing trend over the entire period, whereas the other categories exhibited trends that reversed. Agricultural land continued to grow until 2001 and started to decrease slightly in the 2001-2014 period. A decreasing tendency of agricultural land during the last period is attributed to the rapid expansion of urban areas and homestead plantations. Substantial deforestation of natural vegetation cover occurred in the 1972-1989 period and the proportion of degraded and agricultural lands significantly increased during the same period. However, vegetation cover has started to recover since the 1990s, when some of the agricultural, bare, and grasslands have turned into vegetation cover.

More than 72% of the total landscape of the study area exhibited a shift from one category to other categories during the entire study period. The largest changes occurred during 1972-1989. Whereas both swap and net changes occurred in all periods, the changes attributed to swap were much larger than the net changes. The observed periodical fluctuations in the LULC dynamics of the basin will certainly have implications on the hydrological processes and availability of water both at local and basin scales. Continuous changing in LULC of the basin might have influenced the partitioning of incoming rainfall into different hydrological components. On the other hand, the observed recent watershed management interventions in the basin might have considerably reduced surface runoff at the watershed level. The observed high variability in streamflow without significant change in rainfall pattern in chapter 4 could be attributed to a change in land management practices. This explicitly shows that the perceived changes in LULC dynamics of the basin could be the main drivers for the variability of water resources in the basin.

The most important driving forces for the observed changes were found to be rapid population growth and changing government policies. The recent government policy on environmental rehabilitation and its associated implementation programs have significantly contributed to the recovery of vegetation cover in the region. Strengthening such interventions with high-level participation of local farmers is essential for the sustainability of biophysical resources. Apart from the well-being of the ecosystem and environmental stabilization, much effort is needed to convert the improved vegetation cover into sources of income for the local communities to secure the gains made so far and to prevent further deterioration.

The result of this study can have significant implications on the future land and water management of the basin. Improved understanding of LULC change dynamics will allow more reliable information for improved land water resources management in the basin. The observed LULC change patterns are essential for establishing relationships between hydrological processes and catchment characteristics, which may reveal the effects of the widespread implementation of IWSM programs on the upstream-downstream hydrological linkages at larger spatial scales.

Chapter 6

MODELLING HYDROLOGICAL RESPONSE TO LAND MANAGEMENT DYNAMICS IN GEBA CATCHMENT [4]

Identifying a complex relationship between hydrological processes and environmental changes are always indispensable for improved water management. The Geba catchment, headwaters of Upper Tekeze basin is known for its severe land degradation and recent good experiences in integrated watershed management. This study is anticipated to analyse the hydrological response of this dynamic land management change using an integrated approach of hydrologic model, hydrological alteration indicators and Partial Least Square Regression (PLSR). A distributed hydrological model based on Wflow-PCRaster/Python modelling framework was developed to simulate the hydrological response of land use/cover (LULC) maps from 1972, 1989, 2001 and 2014. Hydrological alteration indicators were applied to understand the degree of streamflow alterations. Finally, the contribution of individual LULC type on the hydrological impacts were quantified using PLSR model.

The results show expansion of agricultural and grazing land at the expense of natural vegetation, resulted in increased annual surface runoff by 77% and decreased dry season flow by 30% in the 1980s and 1990s compared to 1970s. However, natural vegetation started to recover from the late 1990s and dry season flows increased by 16%, whilst surface runoff declined by 19%. More pronounced changes in the hydrological responses were noticed at the sub-catchment level, mainly associated with spatial variability of land degradation and rehabilitation. However, the rate of increase in low flows has slightly decreased in the 2010s, suggesting an increase in water withdrawal for irrigation. Fluctuations in hydrological alteration parameters are in agreement with the observed LULC change. The PLSR analysis demonstrated that most LULC types showed a strong association with all hydrological components.

These findings demonstrate that changing water conditions are in fact interlinked to the LULC change. The combined analysis of rainfall-runoff modelling, alteration indicators, and PLSR is very promising to assess the impact of environmental change on the hydrology of complex catchments.

---

[4] Based on Gebremicael, T.G., Mohamed, Y.A., Van der Zaag, P., (2019a). Attributing the hydrological impact of different land use types and their long-term dynamics through combining parsimonious hydrological modelling, alteration analysis and PLSR analysis, *Sci. Total Environ.*, 660, 1155-1167.

6.1 INTRODUCTION

Understanding variability and change of hydrological processes and their implications on water availability is vital for water resource planning and management. Human-induced environmental changes are key factors controlling the variability of streamflow (Hassaballah et al., 2017; Woldesenbet et al., 2017). Excessive pressure on land resources aimed at providing food, water and shelter have resulted in a significant change of land cover which consequently modified the hydrological regimes (Gyamfi et al., 2016; Savenije et al., 2014). Alteration of existing land management practices in a catchment affects the hydrological processes such as infiltration, groundwater recharge, base-flow and surface runoff and consequently the overall water availability in rivers (Kiptala et al., 2013; Hurkmans et al., 2009). The Land Use / Land Cover (LULC) changes significantly influence the timing and magnitude of extreme events (Woldesenbet et al., 2017; Yan et al., 2016). As such, the effect of LULC change on hydrology has continuously drawn the attention of scientific communities to understand the complex relationship between hydrological processes and human-induced environmental changes. However, the heterogeneity of catchment characteristics coupled with limited hydro-climatological data is the major scientific challenge to fully understand such complex relationships (Gashaw et al., 2018; Gyamfi et al., 2016; Tekleab et al., 2014; Li and Sivapalan, 2011).

The Ethiopian highlands, covering 45% of the country, is affected by severe LULC and land degradation problems (Gashaw et al., 2018; Haregeweyn et al., 2014; Nyssen et al., 2014). Because of the high increase of population, a rapid expansion of cultivable lands has significantly reduced land with natural vegetation (Gebremicael et al., 2013; 2018; Haregeweyn et al., 2017; Ariti et al., 2015; Hurni et al., 2005). Zeleke and Hurni (2001) reported an increase of cultivable land by 77% (1957-1995) in the Dembecha watershed of the Blue Nile basin and a decline of forest coverage by 99%. Haregeweyn et al. (2014) showed that agricultural land in the Gilgel Tekeze watershed increased by 15% at the expense of shrubland which decreased by 19% between 1976 and 2003. In contrast, Wondie et al. (2011) revealed that forest coverage in the Semen mountain national park of the Blue Nile basin increased by 33% in 20 years. These changes can modify water resources availability in those catchments as it affects the partitioning of rainfall into different hydrological components (Taniguchi, 2012). A number of studies investigated impacts of LULC change on the hydrology at different spatiotemporal scales (Hassaballah et al., 2017; Woldesenbet et al., 2017; Chen et al., 2016; Tekleab et al., 2014; Gebremicael et al., 2013). However, the results indicated that the impacts are not universal as it depends on the local context of the specific catchment (Chapter 5; Haregeweyn et al., 2014). While some studies show that the conversion of human-modified land cover back to natural vegetation cover reduces surface runoff while enhancing the base-flow (Chen et al., 2016; Haregeweyn et al., 2014), other studies showed that increased natural vegetation cover reduces runoff as more of the incoming rainfall is contributed to canopy interception and evapotranspiration (Wang et al., 2018; Ott and Uhlenbrook, 2004).

The Upper Tekeze headwaters located in Ethiopia, which is one of the sources of the Nile water, is characterized by severe land degradation. The natural vegetation has been replaced by cultivable and grazing lands during the period of the 1960s to the early 1990s (Tesfaye et al., 2017; Belay et al., 2014). Nevertheless, forestation started to recover from the late 1990s due to watershed management interventions (Belay et al., 2014; Nyssen et al., 2010). Local studies revealed that the impact of LULC on the hydrology of the basin is significant. Abraha (2014) reported that the conversion of natural vegetation to agricultural crops in the upper Geba catchment increased surface runoff by 72% and decreased dry season flow by 32% over 1972-2003. In contrast, Bizuneh (2013) found that despite almost all land had been converted into cultivable area, the response of surface runoff and base-flows did not change in the Siluh watershed, located in the same region. The disagreement suggests the impact of LULC on the hydrological processes is site-specific and varies with the catchment scale. As such, it is necessary to investigate the space-time relationship between LULC and hydrological responses to support informed land and water management interventions.

The effect of LULC change on hydrological processes has been studied using ground measurements, hydrological models, multivariate statistics and paired catchment methods (Gashaw et al., 2018; Kiptala et al., 2014; Shi et al., 2013). Hydrological models, ranging from conceptual to fully physically based distributed models have been applied in different regions. These types of models have their own advantages and disadvantages (Savenije, 2010). The fully distributed physical models are appropriate to accurately describe the hydrological process in a complex catchment (Wang et al., 2016; Savenije, 2010). However, the excessive complexity of models (over-parameterization) makes model calibration extremely challenging (Savenije, 2010; Uhlenbrook et al., 2004). The over-parameterization problem is not the primary concern of conceptual models, but they usually fail to reproduce the non-linear dynamics of catchment characteristic which is essential in studying the hydrological response to the dynamics of environmental changes (Sivapalan et al., 2003). Therefore, to avoid over-parameterization and maximize information retrieved from spatial data, this study attempted to develop a parsimonious dynamic distributed model which requires modest calibration. The literature shows that PCRaster/Python programing language are becoming popular tools to develop dynamic and flexible distributed hydrological models, such as Wflow and TOPMODEL (Wang et al., 2016; Karssenberg et al., 2010). The spatially distributed hydrological models have the potential of simulating the impact of human-induced environmental changes. In this study, a spatially distributed hydrological model based on the Wflow-PCRaster/Python modelling framework was developed to simulate the hydrological processes. The Indicators of Hydrological Alteration (IHA) Model (Mathews and Richter, 2007) was applied to assess the degree of streamflow alterations. The contribution of individual LULC changes on the hydrological components was then investigated using a Partial Least Square Regression (PLSR) model (Abdi, 2010). The IHA tool is robust to assess the magnitude of streamflow alterations obtained from the hydrological model whilst the PLSR method is useful to zoom into which LULC is responsible for this alteration.

6.2 DATA AND METHODS

6.2.1 Input datasets

Static data which contain maps that do not change over time and dynamic data containing maps that change over time are the main inputs for the development of the hydrological model. These datasets are explained in detail in the following sections.

Static data

The static datasets of land cover, Digital Elevation Model (DEM), soil type, Local Drainage Direction (LDD) and river maps are required to develop a distributed hydrological model in a Wflow_PCRaster/Python framework environment (Schellekens, 2014). LULC data of the catchment were acquired from our previous study in Chapter 5. In that study, Landsat images of the years 1972, 1989, 2001, and 2014 for the Geba catchment were processed. The years for analysis were selected based on key signs of LULC change (land degradation, land policy changes and availability of the satellite image). An intensive field data, including ground truth (3326 points), interviews and observations, topographic maps, areal photographs and secondary information from the literature were used to validate the land use classification. A hierarchical classification comprised of unsupervised/supervised approaches was performed to identify nine LULC class types in the catchment (Table 5.2). For more information about the classifications and accuracy assessments, refer to Gebremicael et al. (2018). For this study, these LULC types were re-classified into seven classes where the plantation and forest, as well as bareland and urban classes were merged as forest and bareland, respectively.

The initial soil map, available online from the International Soil Reference and Information Centre (ISRIC) SoilsGrid250 (Hengl et al., 2017), was modified based on the study area characteristics. Additionally, detailed soil properties of: texture, bulk density, available water capacity, hydraulic conductivity, saturated hydraulic conductivity, soil depth, particle-size distribution, were obtained from 160 soil samples collected from the Geba catchment. Soil types from the ISRIC map were reclassified into seven major groups and the physical properties of each soil group were enhanced with the result of the 160 soil samples. A 30 m resolution DEM, used to delineate the catchment boundaries and derive the LDD, were obtained from the Shuttle Radar Topographic Mission (SRTM).

Dynamic input datasets

Daily climate data were collected from 16 stations located within and surrounding the catchment (Figure 5.1). These data were provided by the Ethiopian Meteorological Service Agency (NMA). The consistency and quality of these data were checked and screened as summarized in chapter 4 Table 4.1. Observed rainfall data from the gauging stations were spatially interpolated using Kriging method (Oliver & Webster, 1990). Daily potential evapotranspiration (PET) was estimated from availiable climatic data using Hargreaves method which is suitable in catchments with limited climatic data (Hargreaves et al., 1985). Finally, all dynamic and static input maps were projected to WGS-84-UTM-zone-37N and resampled to a resolution of 100m for the model inputs. Hydrological flows of five gauging stations (location is given Figure 5.1) were collected from the Ethiopian Ministry of Water Resources for calibration and validation of the model. Descriptions and quality of these data are presented in Chapter 2, Table 2.1.

6.2.2 Methods

Development of the hydrological model

In this study, a distributed hydrological model based on the Wflow_PCRaster/Python framework was developed to assess the impact of LULC change on streamflow dynamics. Wflow is an open source software developed by the Deltares OpenStreams project which simulates catchment runoff in both limited and rich data environments (Schellekens, 2014). Wflow_sbm model is based on TOPOG hydrological tool described in Vertessy & Elsenbeer (1999). This model was derived from the CQFLOW model (Köhler et al., 2006) and is programmed in the PCRaster-Python environment (Karssenberg et al., 2010b). It was selected in this study for its improved consideration of both infiltration and saturation excess runoff generation processes. A schematized representation of Wflow_sbm is given in Figure 6.1.

The hydrological processes in the model are represented by three main routines. Interception is calculated using Gash model (Gash et al., 1995) which uses PET to drive actual evapotranspiration based on the soil water content and land cover types. They have introduced a simplified model that can be applied at a daily time step. The amount of water needed to completely saturated the canopy (P') is given as

$$P' = \frac{-\bar{R}S}{\bar{E}_w} \ln \left[1 - \frac{\bar{E}_w}{\bar{R}} (1 - P - P_t)^{-1} \right] \tag{6.1}$$

where $\bar{R}$ and $\bar{E}_w$ are the average precipitation intensity on saturated canopy and evaporation from the wet canopy in mm/day, respectively. S is canopy storage capacity (mm), P is free through fall coefficient and P_t is the proportion of rain diverted to stemflow. Interception losses from the stems are calculated for days with $P \geq \frac{S_t}{P_t}$. S_t and P_t are small and neglected in the WFlow_sbm model. The model assumes saturated conditions occur when the hourly rainfall exceeds a certain threshold level. $\bar{R}$ is calculated for all hours when the rainfall exceeds the given threshold to give an estimate of the mean rainfall rate onto a saturated canopy. $\bar{E}_w$ is then calculated using the Rutter model (Schellekens, 2014; Gash et al., 1995).

The Soil Water Storage (SWS) processes that control runoff generation are calculated by the TOPOG_sbm (Vertessy & Elsenbeer, 1999). TOPOG_sbm was specifically designed to simulate fast runoff processes; however, considerable improvement has been made in Wflow_sbm to make it more widely applicable (Schellekens, 2014). The soil in the model is represented by a bucket with a certain depth (Zt), divided into a saturated (S) and unsaturated (U) stores. The top of the saturated store forms a pseudo-water table at depth (Zi) such that the value of (S) at any time is given by Equation.6.2.

$$S = (Z_t - Z_i)(\theta_s - \theta_r)$$ (6.2)

Where θ_s and θ_r are the saturated and residual soil water contents, receptively. The unsaturated store is subdivided into storage (U_s) and deficit (U_d) which are expressed in depth units as

$$U_d = (\theta_s - \theta_r)Z_i\text{-}U \text{ and } U_{s=}U - U_d$$ (6.3)

The saturation deficit (S_d) for the whole soil profile is defined as

$$S_d = (\theta_s - \theta_r)Z_t\text{-}S$$ (6.4)

Infiltrated rainfall enters the unsaturated store first and the transfer of water from saturated to saturated store (st) is controlled by the saturated hydraulic conductivity (K_{sat}) at depth (Z_i) and the ratio of U_s to S_d as shown below

$$st = K_{sat}\frac{U_s}{S_d}$$ (6.5)

This shows that when the saturated deficit is smaller, the rate of water transfer from U to S increases. The saturated hydraulic conductivity declines with soil depth according to:

$$K_{sat} = K_0 e^{(-fz)}$$ (6.6)

Where: K_0 the saturated hydraulic conductivity at the soil surface, f is scaling parameter (m^{-1}). The scaling parameter f is given by

$$f = \frac{\theta_s - \theta_r}{M}$$ (6.7)

Where: M is a model parameter determining the decrease in saturated conductivity with depth (m). The saturated store is drained laterally via subsurface flow according to the following Eq:

$$sf = K_0 \tan(\beta)e^{-S_d/M}$$ (6.8)

Where β element slope is angle (degrees) and sf is calculated subsurface flow (m²/d).

In Wflow_sbm, transpiration is first taken from the saturated store if the roots reach the water table Z_i. If S store does not satisfied the demand, the U store is used next. First the wet root (WR) number (scaling from 1 to 0) is determined using sigmoid functions as

$$WR = 1.0/1.0 + e^{-SN(WT-RD)}$$ (6.9)

Where SN is sharpness parameter which is a negative large number determines if there are a stepwise or more gradual outputs. WT is the level of water table (mm) in the grid cell below the surface. RD is the maximum depth of the roots in mm below the surface. For all values of WT< RD a value of 1 is returned, if they are equal a value of 0.5 is returned, but if WT>WD a value of 0 is returned. The returned value of WR is multiplied by the potential evaporation) to get the transpiration from the saturated part of the soil. The remaining potential evapotranspiration is used to extract water from the unsaturated zone.

To determine the capillary rise, first the K_{sat} is estimated at the water table Z_i; next a potential capillary rise is determined from the minimum of the K_{sat}, the actual transpiration taken from U store, the available water in the saturated store and the deficit of the unsaturated store. Finally, the potential rise is scaled using the distance between the roots and the water table using

$$CS = CSF/(CSF + z_i - RD) \qquad (6.10)$$

Where CF is scaling factor to multiply the potential capillary rise, CSF is a model parameter. If the roots reach the WT (RD > Z_i), CS is set to zero, thus setting the capillary rise to 0.

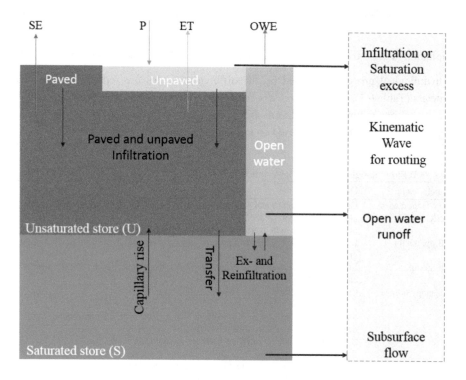

Figure 6.1: Schematization of the different processes and fluxes in wflow_sbm model (modified from Schellekens, 2014). P = precipitation; SE = soil evaporation; ET = Evapotranspiration, OWE = open water evaporation.

The river drainage and overland flows are modelled using kinematic wave routing. Rainfall and evaporation in the saturated canopy are calculated for each event to estimate the average rainfall and evaporation from the wet canopy. The remaining water infiltrates into the soil and when the rain falls on partially saturated soil, it directly contributes to surface runoff. At the same time, part of the soil water is taken by evapotranspiration. The infiltrated water exchanges between the unsaturated stores (U) and saturated store (S) of the soil (Figure 6.1). The soil in Wflow_sbm is considered as a simple bucket model which assumes an exponential decay of the saturated hydraulic conductivity (Ksat) depending on the depth (Schellekens, 2014). The soil depth of the different land cover types in the model is identified and scaled using a topographic wetness index (Vertessy & Elsenbeer, 1999). As the model is fully distributed, the runoff is calculated for each grid cell with the total depth of the cell being divided into saturated and unsaturated zones (Figure 6.1). Darcy's equation is applied in the model to simulate lateral flow from the saturated zone. The total runoff from a given catchment is the sum of surface runoff and lateral flow which is routed from the river network as discharge using kinematic wave routing. Deep percolation to the groundwater table (countribution to groundwater recharge) is calculated outside the model from the water balance equation shown in Equation 6.11.

Precipitation - Evapo-transpiration – river discharge -Deep Percolation = ΔStorage   (6.11)

The hydrological process described by the different modules are represented by 19 main parameters (Table 6.1). These parameters are linked to the model by a PCRaster look-up table that contains four columns. The columns are used to identify the land use type, sub-catchment, soil type and the values assigned based on the first three columns, respectively.

Table 6.1: WFlow model parameter's description

Parameter name	Description	Unit
CanopyGapFraction	Gash interception model parameter, the free throughfall coefficient	[–]
EoverR	Gash interception model parameter. Ratio of average wet canopy evaporation rate over average precipitation rate	[–]
InfiltCapPath.tbl	Infiltration capacity of the compacted soil (or paved area) fraction of each gridcell	[mm/day]
InfiltCapSoil	Infiltration capacity of the non-compacted soil fraction (unpaved area) of each gridcell	[mm/day]
MaxCanopyStorage	Canopy storage used in the Gash interception model	[mm]
K_{sat}	Saturated conductivity of the store at the surface. The M parameter determines how this decreases with depth.	[mm]
M	Soil parameter determining the decrease in saturated conductivity with depth	[m]
N	Manning N parameter for the kinematic wave function	[S/m$^{-1/3}$]
N_River	Manning's parameter for cells marked as river.	[S/m$^{-1/3}$]
PathFrac	Fraction of compacted area per grid cell.	[–]
RootingDepth	Rooting depth of the vegetation	[mm]
SoilMinThickness	Minimum depth of the soil	[mm]
SoilThickness	Maximum depth of the soil	[mm]
CapScale	Scaling factor in the capillary rise calculations	[mm/day]
thetaR	Residual water content.	[–]
thetaS	Water content at saturation (porosity).	[–]
Beta	Element slope angle	[degree]
rootdistpar	Sharpness parameter determine how roots are liked to water table	[mm]
RunoffGeneratingGW Perc	Fraction of the soil depth that contributes to subcell runoff	[–]

Model calibration and validation

The developed hydrological model was calibrated at several locations of the catchment. First, it was calibrated and validated at Geba1 with a catchment area of 4590 km^2 and subsequently at four smaller catchments (Geba2 2440 km^2; Siluh 960 km^2; Genfel 730 km^2; and Ilala, 340 km^2) (Figure 5.1). Calibration of the model started with a selection of parameters and their initial values from the literature and our own field observations and laboratory analysis. Next, parameter values were manually adjusted until maximum concordance between observed and predicted streamflow occurred. The model was then evaluated using different objective functions to verify whether the predicted and observed streamflow agreed.

The model was calibrated and validated using LULC of the years 1972 and 2014, separately. Observed datasets of 1972-1974 and 1975-1977 were used for calibration and validation of the model using the LULC map of 1972. Similarly, datasets of 2010-2012 and 2013-2015 were used for the calibration and validation of the model using LULC of 2014. The two calibrated models were used to simulate the hydrological responses for each of the four LULC maps (1972, 1989, 2001 and 2014). These models and input datasets are summarized in Table 6.2. Nash–Sutcliffe Efficiency (NSE) and Percent Bias (PBIAS) statistical indices were applied to evaluate the performance of the model. Detailed descriptions of these indices are given in chapter 3 Table 3.2.

Table 6.2: Summary of the different developed models and their input datasets

Model name	Model parameters	Input datasets	Year of LULC	Remarks
F_1972			1972	
F_1989	P_1972	1971-1977	1989	Forward modelling
F_2001			2001	
F_2014			2014	
R_2014			2014	
R_2001	P_2014	2009-2015	2001	Reverse modelling
R_1989			1989	
R_1972			1972	

Impact modelling approach

The two calibrated models (1972, and 2014) were applied to the classified LULC maps of 1972, 1989, 2001 and 2014 to assess the impact of LULC change on the hydrology which results in 8 model outputs (Table 6.2). These maps were used separately to run the model by keeping other model inputs constant. The "fixing-changing method" that is changing LULC maps while keeping model parameters from both calibrated models (e.g., P_1972 and P_2014) and other input datasets (hydrology, climate, soil and DEM) constant was used to investigate the impacts. This approach has been widely used in a number of studies to assess the impact of LULC change on hydrological response (Gashaw et al., 2018; Woldesenbet et al., 2017; Gyamfi et al., 2016; Yan et al., 2016). As streamflow data is not available in the 1980s and early 1990s, the "fixing-change simulation method" is suitable to simulate the hydrological response attributed to LULC change in the 1980s.

First, the calibrated model parameters using the LULC map of 2014 and other input datasets were used to simulate the hydrological response of 2014, 2001, 1989 and 1972 LULC maps. Second, a similar procedure was applied to see the hydrological responses of 1972, 1989, 2001 and 2014 LULC maps using the calibrated model parameters with the 1972 LULC map. As summarized in Table 6. 2, the same model parameters and input datasets from 2014 and 1972 models were used to simulate the hydrological responses of 2014, 2001, 1989 and 1972 LULC maps in the reverse modelling and forward approaches, respectively. Applying both reverse and forward modelling approaches is essential to minimize input data uncertainties during calibration and validation processes (Yu et al., 2018). Due to the observed dynamic change in LULC (Table 5.1), the value of model parameters is expected to vary during the study period. Hence, employing both reverse and forward modelling approaches is essential to conduct an in-depth analysis of the hydrological response to LULC change dynamics from both directions.

Application of Indicators of Hydrological Alterations (IHA)

The change of streamflow dynamics caused by the change in LULC, as simulated by the hydrological models, was also quantified by Indicators of Hydrological Alterations (IHA). The IHA software developed by the US Nature of Conservancy (Mathews and Richter, 2007) were used to detect the hydrological fluctuations in the catchment. Characterizing these hydrologic parameters is essential to understand the variation of hydrological systems before and after environmental changes (Hassaballah et al., 2017; Saraiva-Okello et al., 2015). 33 IHA parameters were considered to characterize the (simulated) flow variations between 1972-1989, 1989-2001, and 2001-2014, including monthly flow condition, magnitude and timing of extreme flows, flow pulses and rates of change.

Partial least square regression (PLSR) analysis

Analysing hydrological responses and change using hydrological simulation and IHA analysis cannot reveal the contribution of each LULC type to hydrological change. Combined use of the hydrologic model, IHA and PLSR could be a viable approach to scrutinize the impact and contribution of each LULC change on the catchment hydrological responses. The pair-wise Pearson correlation combined with the PLSR model (Abdi, 2010) was applied to further investigate the relationship between individual LULC types and each hydrological component.

This approach is essential to ascertain whether the observed change in LULC was large enough to cause the change in streamflow dynamics. The relation between each LULC type and hydrological components was computed using the pair-wise Pearson correlation whilst the contribution of their change to the streamflow was quantified using the PLSR model. The PLSR is a robust multivariate regression technique that is appropriate when the response (dependent variables) exhibit collinearity with many predictors (independent) variables (Woldesenbet et al., 2017; Shi et al., 2013). It combines features from principal component analysis and multiple regressions that is appropriate when predictors exhibit multicollinearity (Yan et al., 2016). In this study, the independent variables are the different LULC types (Table 5.1) while the dependent variables are hydrological components (total runoff, wet and dry season flow, actual evapotranspiration (AET) and SWS). Detailed information on PLSR algorithms can be found in the literature (Shi et al., 2013; Abdi, 2010) and hence only a brief description is given here. The PLSR is a linear model specifies the relationship between a response variable, Y, and a set of predictor variables X's given in Equation 6.12.

$$Y = a_0 + a_1X_1 + a_2X_2 + a_3X_3 + \cdots + a_iX_i \qquad (6.12)$$

where Y is the response variable, a_0 is the intercept, X is the independents variables from 1 to i, and b is the coefficients of the x variables.

An interesting feature of the PLSR model is that the relationship between the independent and dependent variables can be inferred from weights (w*) and regression of each independent variable in the most explanatory components (Abdi, 2010). This is essential in order to identify which LULC is strongly associated with the streamflow. The quality and strength of the model are measured by the proportion of variance in the matrix of independent variables used in the model (R2x), the proportion of the variance in the matrix of dependent variables explained by the model (R2y) and cumulative goodness of prediction within a given number of factors (Q2cum). Values of R2x, R2y > 0.5 and Q2cum > 0.097 are considered as a good predictive ability of the model (Tenenhaus, 1998). Cross-validation was used to determine the number of significant PLSR components. Detailed information on the calculations of these indices are explained in Shi et al. (2013) and Yan et al. (2013). The importance of predictors of both independent and dependent variables of the PLSR modelling is given by the Variable Influence Projection (VIP). Predictors with higher values of VIP better explain the consequence of the independent on the dependent variables. As a rule of thumb, VIP > 1 is statistically significant to explain the dependent variables (Yan et al., 2017). Weight (w*) coefficients in the PLSR model describe the direction and strength of contributions from each independent variable (Yan et al., 2016). Small values of VIP and W* reveal that the variable is not relevant to explain the dependent variable and can be excluded from the model. To infer if the samples are given from a normally distributed population or not, normality was checked using the Shapiro and Wilk (1965) normality test. The PLSR modelling and other statistical analysis including the multicollinearity of predictors were performed with SPSS software (Carver and Nash, 2009) and XLSTAT tool (www.XLSTAT.com).

6.3 RESULT AND DISCUSSION

6.3.1 Calibration and validation of the Wflow hydrological model

The developed models were calibrated and validated using the LULC of 2014 and 1972. The 2014 model has been calibrated and validated at five locations, while two locations were used for the 1972 model, respectively. Forward modelling was done at only two stations due to the absence of streamflow data in the 1970s at three locations. The simulated and observed streamflow in the calibration and validation periods using the 2014 and 1972 models are given in Figure 6.2 and Figure 6.3, respectively.

Prior to calibration, one year data (2009) were used to initialize the model conditions and identify the most sensitive parameters. Model parameters such as saturated hydraulic conductivity (Ksat), residual water content (thetaR), CanopyGapFraction, M parameter (controls decay of hydraulic conductivity with depth), and Manning coefficient (N) were the most important parameters controlling outflow. Optimized model parameter values after the calibration processes in different stations are summarized in Appendix D (Table D-1 to D-5). Parameter values varied from sub-catchment to sub-catchment within the same calibration period and from time (the 1970s) to time (the 2010s). For example, the average value of CanopyGapFraction for all LULC types is higher in Geba2 (0.28) and Siluh (0.26) than in Geba1 (0.2). Similarly, the value of this parameter significantly reduced from the 1970s (0.52) to 2010s (0.25).

As presented in Figure 6.2 and Figure 6.3, the models were able to simulate the observed flow consistently in all gauging stations during the 2010s and 1970s calibration processes.

The performance indices for the daily calibration and validation are listed in Table 6.3. The value of NSE during calibration and validation is greater than 0.6 with PBIAS around ±25% in three stations during the 2010s and in both stations during the 1970s comparisons (Table 6.3). This suggests a very good model performance (Moriasi et al., 2007). The model performed relatively less in Genfel and Illala catchments with NSE of less than 0.6 and higher PBIAS during validation (Table 6.3 and Figure 6.2). The positive value of PBIAS shows the tendency of the model to consistently overestimate the streamflow across all gauging stations (Figure 6.2 and 6.3). For example, the streamflow was overestimated by 9.6%, 13.2% and 11.2% during calibration and 11.8%, 18% and 14.3% during validation in Geba1, Geba2 and Siluh stations, respectively. In contrast, peak flow in Geba1 and Siluh were slightly underestimated during 2010 and 2011, respectively. Such overestimation and slight underestimation could also be attributed to the interpolation of the sparse and unevenly distributed rain gauges over the complex terrains of the catchment.

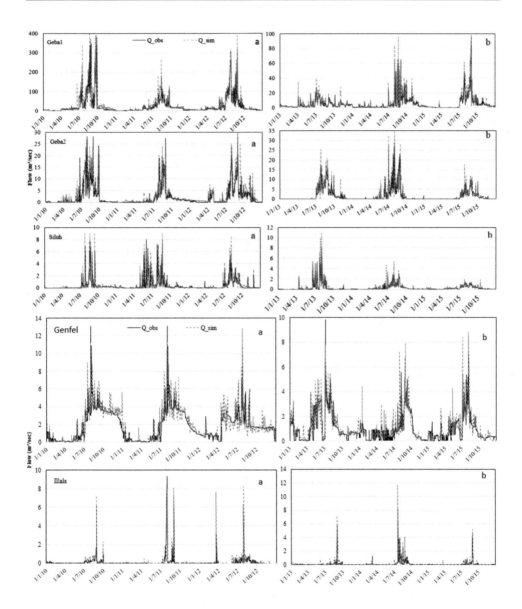

Figure 6.2: *Calibration (a) and validation (b) of Wflow model using LULC of 2014 at five locations of the catchment*

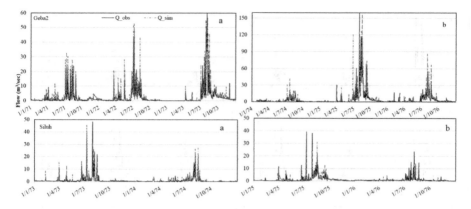

Figure 6.3: Calibration (a) and validation (b) of Wflow model using LULC of 1972 in two locations of the catchment

Comparing the different sub-catchments, the performance of the model slightly improved at the downstream stations (Table 6.3). The likely reason is that some of the errors at a smaller scale counter-balanced each other when combined at downstream stations. Generally, the consistency of simulated and observed hydrographs and statistical indices indicate that the model was able to describe the daily streamflow of the catchment. Thus, the calibrated models in the 2010s and 1970s were applied to simulate the impact of LULC change in the catchment.

Table 6.3: Performance criteria of the model calibration and validation at daily scale

Catchment	2014 model (Reverse modelling)				1972 model (Forward modelling)			
	Calibration		Validation		Calibration		Validation	
	NSE	PBIAS (%)	NSE	PBIAS (%)	NSE	PBIAS (%)	NSE	PBIAS (%)
Geba1	0.83	12.35	0.81	10.73				
Geba2	0.76	19.22	0.83	10.73	0.77	12.27	0.75	14.46
Siluh	0.86	15.14	0.84	21.83	0.71	24.39	0.67	18.07
Genfel	0.69	61.91	0.58	14.29				
Illala	0.55	23.94	0.49	27.72				

6.3.2 Streamflow responses to LULC changes

Figure 6.4 and 6.5 shows streamflow of the sub-catchments simulated from LULC maps of different periods. Comparison between the hydrographs obtained from each LULC map indicated that the streamflow is significantly affected by the observed change in LULC in all sub-catchments. For example, the peak flows from the LULC map of 1989 were higher than from the remaining maps during the reverse (Figure 6.4) and forward (Figure 6.5) modelling approaches. Although the magnitude of peak flow from 2001 and 2014 maps is still higher than the 1972 map, it has slightly decreased compared to the 1989 LULC. In contrast, the low flows during the dry months have considerably decreased from 1972 to 1989 and started to increase from 1989 to 2001. The rate of increase of the low flows from 2001 to 2014 halted in most catchments. The results from the reverse modelling approach (Figure 6.4) is in agreement with the forward modelling approach (Figure 6.5) where in both cases similar patterns of change in peak and dry season flows were observed.

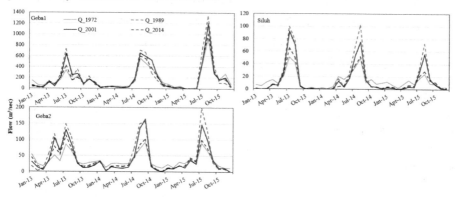

Figure 6.4: Comparison of simulated streamflow at monthly time steps in multiple sub-catchments and different LULC (1972, 1989, 2001 and 2014) using reverse modelling approaches (2014 model).

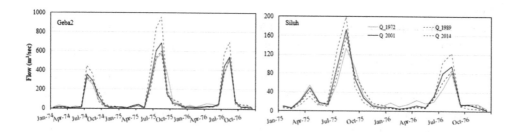

Figure 6.5: Comparison of simulated streamflow in two sub-catchments and different LULC (1972, 1989, 2001 and 2014) using forward modelling approaches (1972 model).

During the reverse modelling approach, the total runoff from the main outlet (Geba 1) has increased by 38% from 1972-1989 and then decreased by 15% and 34% using 2001 and 2014 maps, respectively (Table 6.4). The wet season flow exhibited a similar pattern to total runoff. Contrasting to the annual and wet season flows, the dry season flow decreased by 23% and increased by 17% during the 1972-1989 and 1989-2001 periods, respectively. However, a decrease of 25% was noticed using 2014 LULC (Table 6.4). As with the SWS and AET fluxes, the average value of AET over the whole catchment decreased by 18% in 1989 compared to 1972 and then increased by 13% and 15% in 2001 and 2014, respectively. Similarly, SWS in soil and contribution to groundwater recharge decreased in the first period and increased during the second and third periods (Table 6.4). A summary of the results for the two remaining sub-catchments (Genfel and Ilala) during reverse and forward modelling (Geba2 and Siluh) are given in Appendix D (Tables D-6 and D-7). The long-term change pattern of each hydrological component corresponding to the observed LULC change periods is also illustrated in Figure 6.6. The value of each hydrological component is normalized by the mean annual rainfall.

Table 6.4: Mean annual (2013-2015) hydrological fluxes in (mm/year) of each LULC maps from the reverse modelling approach (2014 model) at different sub-catchments

	Land use/cover				Change in fluxes (%)			
Water balance	1972	1989	2001	2014	1972-1989	1989-2001	2001-2014	1972-2014
Geba 1								
Annual rainfall (mm/year)	600	600	600	600	0	0	0	0
AET (mm/year)	407	332	376	434	-18	13.3	13	7
Deep Percolation (mm/year)	55	36	38	45	-19	2	7	-10
SWS (mm/year)	20	6	12	21	-12	6	14	1
Annual flow (mm/year)	181	250	213	141	38	-14.8	-51	-22
Wet season flow (mm/year)	111	196	150	94	77	-23.5	-60	-15
Dry season flow (mm/year)	70	54	63	47	-30	16.7	-34	-33
Runoff coefficient	0.30	0.42	0.36	0.24	40	-14.3	-50	-20
Wet season /annual flow	0.61	0.78	0.70	0.67	28	-10.3	-4	10
Dry season /annual flow	0.39	0.22	0.30	0.33	-44	36.4	9	-15
Geba 2								
Annual rainfall (mm/year)	550	550	550	550	0	0	0	0
AET (mm/year)	328	270	329	351	-18	21.9	6	7
Deep Percolation (mm/year)	63	45	59	67	-18	14	8	4
SWS (mm/year)	44	11	20	26	-33	9	6	-18
Annual flow (mm/year)	123	224	150	103	82	-33.0	-46	-16
Wet season flow (mm/year)	72	166	102	64	131	-38.6	-59	-11
Dry season flow (mm/year)	58	51	47	37	-12	-7.8	-27	-36
Runoff coefficient	0.22	0.41	0.27	0.19	86	-34.1	-42	-14
Wet season/annual flow	0.59	0.74	0.68	0.62	25	-8.1	-10	5
Dry season/annual flow	0.41	0.26	0.31	0.36	-37	19.2	14	-12
Siluh								
Annual rainfall (mm/year)	540	540	540	540	0	0	0	0
AET (mm/year)	369	313	355	388	-15	13.4	9	5
Deep Percolation (mm/year)	70	43	80	90	-27	37	10	20
SWS (mm/year)	54	32	29	42	-21	-3	13	-12
Annual flow (mm/year)	92	165	108	72	79	-34.5	-50	-22
Wet season flow (mm/year)	49	142	80	34	190	-43.7	-135	-31
Dry season flow (mm/year)	33	23	28	16	-30	21.7	-75	-52
Runoff coefficient	0.17	0.31	0.20	0.13	82	-35.5	-54	-24

| Wet season /annual flow | 0.53 | 0.86 | 0.74 | 0.47 | 62 | -14.0 | -57 | -11 |
| Dry season /annual flow | 0.36 | 0.14 | 0.26 | 0.22 | -61 | 85.7 | -18 | -39 |

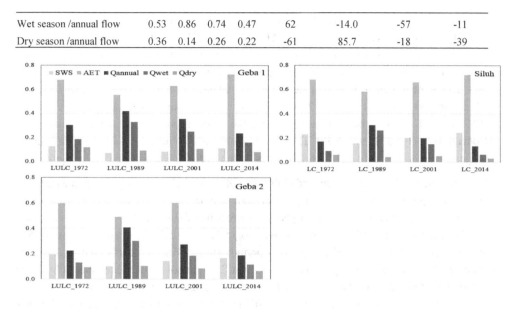

Figure 6.6: Long-term change pattern of each hydrological components corresponding to the observed land use/cover change for the last 44 years.

At the sub-catchment level, more pronounced changes of hydrological flows could be noted. For example, the observed total surface runoff in 1972 at Geba2 increased by more than 82% in 1989 while SWS and AET decreased by more than 49 and 18%, respectively. Similarly, in Siluh sub-catchment, AET and SWS declined by 32% and 15%, whereas the total runoff increased by more than 100% for the same period. Such changes are mainly associated with the uneven spatial distribution of land degradation over the catchment. Despite that the absolute values from the forward modelling are not the same as the values from the reverse modelling due to the differences in climatic inputs, the relative magnitude of changes in the hydrological components are close to each other.

6.3.3 Hydrological alteration trends in response to the observed LULC changes

The hydrological response observed from the hydrological model in section 6.3.2 was further quantified by the IHA method, using the results of the 2014 model. The IHA analysis based on model results indicates that there has been a continuous alteration of the hydrological variables in the Geba catchment after the occurrence of LULC changes between the different periods (Figure 6.7). The magnitude of the median monthly flow between 1972- 1989 increased by an average of 52% during the wet months while it decreased by 49% in the dry months (Figure 6.7a).

The reverse pattern was observed in the remaining analysis period, where the average median monthly flow of the wet months decreased by 17% and 22% and the dry months increased by 30% and 29% in the 1989-2001 and 2001-2014, respectively. With an increase of agricultural land by 42% and a decrease of natural vegetation cover by 36%, the average median monthly flow during the wet and dry months has increased and decreased by 4% and 23%, respectively. All parameters increased from 1972 to 1989 except the 7-day maxima, which decreased from 1989 to 2001. During the 2001-2014 period, the median value of all maxima parameters moderately declined (Figure 6.7b). The observed changes on the annual maxima and minima suggests that the influence of LULC dynamics on the hydrological processes were significant.

The frequency and duration of low and high flow pulses were also investigated for the response of each map (Figure 6.7c). The annual pulse count and their duration increased in the first period and consistently decreased in the later periods. Increasing of pulses below and above the given threshold in the first period shows that the hydrological responses in the catchment were flashier in the 1980s and 1990s. In the latter two periods (the 2000s and 2010s) the peak runoff hydrographs declined in most part of the catchment. Most importantly, the number of high and low pulses are related to the rise and fall rate which give a good understanding of how the streamflow response to catchment characteristic is increasing/ decreasing. Figure 6.7d shows the trend of median daily flow rising and falling rates resultant from each LULC map. Like the high and low pulse pattern, the median rise rate (positive differences between two consecutive daily values) has increased from 1.1 m^3/sec in 1972 to 1.6 m^3/sec in the 1989 land use and then started to decrease by 0.4 and 0.65 m^3/sec from 1989 to 2001 and to 2014, respectively. In the same way, the median fall rate (negative differences) increased from 0.2 to 0.6 m^3/sec in 1972 to 1989 and then decreased by 0.1 m^3/sec in the latter two periods (2001 & 2014).

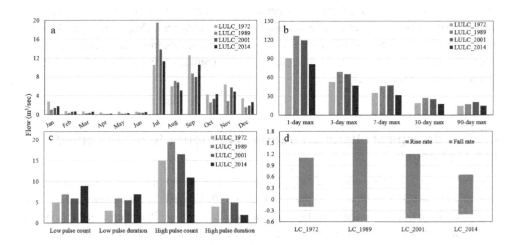

Figure 6.7: Comparison of hydrological responses for each of the four LULC maps using hydrological alteration parameters: (a) Magnitude of median monthly flow; (b) Magnitude of median annual maxima flows; (c) Frequency and duration of high and low pulses; and (d) Rise and fall rate of flows.

6.3.4 Hydrological impacts of individual land use/land cover changes

A preliminary analysis using a pair-wise correlation matrix indicated that most LULC types have a strong association with the change in hydrological components (Table 6.5). Natural vegetation cover including wood and bushland have a significant negative correlation with annual and wet season flows, but a strong positive correlation with dry season flow, SWS and AET in the catchment. In contrast, the expansion of agricultural and bareland showed a significant positive correlation with annual and wet season flows and a significant negative correlation with SWS and dry season flow (Table 6.5). Although statistically not significant, the grassland and water body classes also showed a positive and negative effect on the different hydrological components. The significant correlation between all except grassland and water classes indicates that changes in these land use categories were the main driving force for the observed change in hydrological regimes.

Table 6.5: Pearson correlation matrices for the change in LULC types and different hydrological components (streamflow, Soil water storage, and Evapotranspiration) in Geba 1 during the reverse modelling approach and for the entire period (1972-2014). Bold values indicate a significant relationship at a level of $P<0.05$.

Variables	AGRI	WOOD	FORE	BARE	WATE	BUSH	GRAS	Annual	Wet	Dry	SWS	AET
AGRI	1											
WOOD	-0.96	**1.00**										
FORE	-0.84	**0.95**	**1.00**									
BARE	0.16	**-0.96**	-0.64	**1.00**								
WATE	0.69	-0.48	-0.20	-0.52	**1.00**							
BUSH	**-0.98**	**0.96**	**0.96**	**-0.98**	0.48	**1.00**						
GRAS	-0.61	0.77	0.94	-0.82	0.15	**0.98**	**1.00**					
Annual	**0.95**	-0.95	-0.57	**0.98**	-0.69	**-0.97**	-0.82	**1.00**				
Wet	**0.97**	**-0.96**	-0.68	**0.98**	-0.56	**-0.97**	-0.87	**0.98**	**1.00**			
Dry	**-0.98**	**0.96**	0.41	**-0.98**	-0.63	**0.98**	0.17	0.21	-0.01	**1.00**		
SWS	**-0.98**	**0.99**	**0.96**	**-0.98**	0.15	**0.95**	**0.98**	-0.82	**-0.94**	0.30	**1.00**	
AET	-0.49	**0.97**	**0.96**	**-0.96**	0.64	**0.97**	0.84	**-0.99**	**-1.00**	-0.08	0.86	1

The obtained values of R2x, R2y and Q2cum are above 0.5 which suggests a good predictive capacity of the models. A summary of three PLSR models constructed separately for streamflow, AET and SWS during the reverse modelling exercises are presented in Tables 6.6 and 6.7. The first components for streamflow, AET, and SWS accounted for 58.2, 75.4, and 97.3% of the total variance, respectively (Table 6.6). The addition of the second component improved the prediction error in streamflow and AET, which cumulatively explained 84.6 and 97.9% of the total variance, respectively. Further addition of third model components for the streamflow explained the total variance by 97.8% (Table 6.6). This notwithstanding, the addition of the second model for SWS and third PLSR model component for AET did not significantly improve the predictive capacity of the models. Prediction errors decrease with an increase in the number of components; however, adding more components can also lead to a greater prediction error, which means that the later added components may not strongly correlate with the residuals of predicted variables (Yan et al., 2016).

Table 6.7 presents the summary of weights and VIP of individual LULC classes. Although the weights are important to show the strength and direction of impacts, a comprehensive demonstration of the relative importance of the predictors can be explained by their VIP values. In the case of streamflow, the highest VIP value was obtained from agriculture, followed by bareland and bushland (Table 6.7). The streamflow appeared to increase with the expansion of agricultural and bareland, whereas the negative values of weight indicate the natural vegetation cover, including bush, forest and woodlands contributed to a decreasing trend of streamflow in the catchment. A lower VIP value for grassland (0.79) and water body (0.81) suggests the contribution of these LULC types on the hydrological change is not significant compared to the others. Similarly, LULC types most closely associated with AET fluxes (shown by high VIP values) were bareland (1.30), bushland (1.30), and forest (1.08). However, their contribution to AET is in the opposite direction, where the AET decreases significantly with increasing of bareland but bush and forest coverages encouraged an increment (Table 6.7). With regard to soil water (SWS), all but agriculture and water body obtained a VIP of greater than one and except bareland all LULC types contributed to an increase in SWS. In summary, the PLSR model identified the main land use dynamics that have affected the change in hydrological components of the catchment. PLSR model results from the forward modelling approach show a similar pattern to the reverse modelling (see Table D-8 and D-9).

Table 6.6: Summary of the PLSR models of streamflow, AET and SWS hydrological components in Geba 1 during the entire study period during the reverse modelling approach (1972-2014).

Response Y	R^2x	R^2y	Q^2	Component	% of explained variability in y	Cumulative % of explained variability in y	RMSEcv (mm)	Q^2_{cum}
Streamflow	0.83	0.87	0.76	1	58.2	58.2	11.3	0.87
(annual. wet &				2	26.4	84.6	10.2	0.83
dry season				**3**	**13.2**	**97.8**	**8.9**	**0.94**
flows)								
AET	0.90	0.89	0.98	1	75.4	75.4	18.3	0.87
				2	**22.5**	**97.9**	**14.8**	**0.96**
SWS	0.67	0.98	0.87	1	97.3	97.3	3.6	0.87

Table 6.7: Variable importance of the projection values (VIP) and PLSR for the hydrological components of Geba 1during the entire study period during the reverse modelling approach (1972-2014)

PLSR	Streamflow (Annual, wet and dry)				AET			SWS	
Predictors	VIP	W*(1)	W*(2)	W*(3)	VIP	W*(1)	W*(2)	VIP	W*(1)
AGRI	1.37	0.08	**0.48**	0.26	0.68	-0.05	0.25	0.93	**-0.38**
WOOD	0.90	-0.23	**-0.47**	**-0.42**	0.82	**0.40**	0.20	1.01	**0.38**
FORE	0.93	**-0.35**	0.20	-0.19	1.07	**0.63**	0.25	1.16	**0.44**
BARE	1.29	**0.51**	0.15	**-0.68**	1.30	**-0.51**	0.28	1.11	**-0.42**
WATE	0.81	-0.21	-0.28	-0.18	0.68	0.24	0.16	0.19	0.07
BUSH	1.12	**-0.51**	-0.30	-0.39	1.30	**0.52**	0.21	1.16	**0.44**
GRAS	0.79	-0.26	-0.05	**-0.33**	0.63	0.25	-0.03	1.23	**0.47**

W >0.3 and <-0.3 suggests PLSR components are mainly weighted on the corresponding variables. Negative and positive values show direction of the regressions*

The results from both forward and reverse hydrological modelling, IHA analysis and PLSR model demonstrated that the hydrological response in the Geba catchment has been significantly affected by the observed dynamic LULC change in the last four decades. Over the whole period (1972-2014) of analysis, the total runoff, wet and dry season flows and SWS has decreased whereas average AET over the catchment has increased. This is due to the overall net decrease in natural vegetation cover and the net increase in the agricultural land (Table 4.1). An increase in peak flow and a decrease in low flow during 1972-1989 is attributed to the rapid expansion of cultivable and grazing lands at the expense of natural vegetation cover between the mid-1970s to the end of 1990s. The decline in natural vegetation cover during this period (Chapter 5) contributed to low infiltration and canopy interception so that the incoming rainfall was converted into surface runoff.

The decrease in surface runoff and increase in SWS and AET in the 2000s and 2010s results from the notable improvement of natural vegetation cover in the 2001 and 2014 LULC maps. The detected increment in natural vegetation cover during the last two periods influenced the partitioning of incoming rainfall to contribute more evapotranspiration and enhance the infiltration capacity and resulted in reduced surface runoff. However, the rate of increase in low flows stalled in the 2010s in most sub-catchments which might be explained by an increase in water withdrawals for irrigation. Several local studies (e.g., Gebremeskel et al., 2018; Kifle and Gebretsadkan, 2016; Nyssen et al, 2010; Alemayehu et al., 2009) reported that irrigated agriculture in the catchment has increased by more than 280% from 2006 to 2015. For example, irrigated agriculture in Genfel and Agulae sub-catchments (Figure 4.1) has increased from 83 and 143 hectares in 2006 to more than 643 and 946 hectares in 2015, respectively.

Model parameters value for two different periods (model in the 1970s and 2010s) were also compared to infer the possible causes of changes. The differences between the values of the two model parameters (see example in Table D-2 and D-4) is attributed to the modification of the basin physical characteristics. For example, parameter values of Ksat, M parameter and Gash interception model (EoverR) increased from 1972 to 2014, while the values of CanopyGapFraction decreased for the same period. This suggests that with an increase in natural vegetation cover more water contributed to infiltration and evapotranspiration instead of going to surface runoff. Change in model parameters between the two periods indicates a change in catchment characteristic response behaviour (Gebremicael et al., 2013; Seibert and McDonnell, 2010).

The findings of IHA and PLSR analysis are consistent with the result from the hydrologic model that expansion of agriculture and grazing land in the last four decades contributed to an increase of surface runoff and a decrease of AET and SWS in the catchment. Alteration of monthly flows during the dry and wet seasons are attributed to the dynamic LULC change of the catchment. The IHA analysis result showed that the magnitude of median annual flow maxima and minima were significantly affected by all LULC maps. The direction of change in flow rise and fall rates was in agreement with the observed LULC changes which implies that the changes in the rate and frequency of water conditions are linked to the LULC change of the catchment. Surface runoff generation showed a strong negative correlation with forest, wood and bushlands whilst a strong positive relationship occurred between dry season flow and these LULC types. The observed dynamic change in these LULC types during the different analysis periods inversely affected the infiltration capacity of the soil and subsequently overland flow to the streams. It is also reported in several studies (e.g., Gashaw et al., 2018; Weldesenbet et al., 2017; Yan et al., 2016) that decrease in vegetation cover contributed to an increase in surface runoff and decrease in dry season flows.

Our findings are in agreement with previous local and neighbouring basin studies. A similar decline in the dry season and increase in wet season flows was reported in the Geba2 sub-catchment. Abraha (2014) showed that the conversion of natural vegetation to agricultural crops in the upper Geba catchment (Geba2) increased surface runoff by 72% and decreased dry season flow by 32% over 1972-2003. Nyssen et al. (2010) showed that the surface runoff volume significantly reduced after catchment management interventions in My ZigZag. Descheemaeker et al. (2006b) found a reduction of surface runoff by 80% after the restoration of vegetation cover in the same watershed. Similarly, Negusse et al. (2013) found that the availability of groundwater in Arbiha Weatsbiha watershed of Genfel sub-catchment increased by more than ten times from 1993 to 2013. Bizuneh (2013), in contrast, found that despite almost all land had been converted into a cultivable area, surface runoff and base-flows did not change in the Siluh watershed. It is not clear why this finding disagrees with results of all other studies in this basin and other neighbouring basins. What all referred studies share is that these were based either on experimental plots or at very small watershed levels, the findings of which are difficult to extrapolate to catchment scale. The approach in this study adopted the catchment scale and uniquely integrates hydrological simulations to identify the hydrological response of land management dynamics, detect the magnitude of the fluctuations of the simulated streamflow and then identify the contribution of each LULC type on the change in streamflow.

This approach has explicitly demonstrated the impact of land management interventions on the hydrology with a better understanding at different spatial scales. The results are also consistent with several studies in the neighbouring basins which reported an increasing trend of wet season flows while the dry season flows decreased due to the conversion of natural vegetation cover into agricultural and grazing lands (Gashaw et al., 2018; Woldesenbet et al., 2017; Haregeweyn et al., 2014; Tekleab et al., 2014; Gebremicael et al., 2013). For example, Tekleab et al. (2014) and Gebremicael et al. (2013) reported that the conversion of natural vegetation cover into agricultural and bareland in the Upper Blue Nile basin has caused an increase of surface runoff and a decrease of base-flows up to 75% and 50%, respectively.

It is essential to point out one major limitation of this study: it did not quantify actual water abstractions over the study period, and therefore the observed flows at gauging stations were not naturalized. As water abstractions during the low flow season have likely significantly increased since around 2010, this must have influenced some hydrological model parameters. Naturalizing the streamflow from abstraction can improve the model and subsequently the results from this approach.

6.4 CONCLUSION AND RECOMMENDATIONS

The investigation of the effect of LULC change to the hydrological flow from the Geba catchment, over the period 1972 to 2014, has shown that, the expansion of agricultural and grazing land at the expense of natural vegetation cover during the period 1972 to 1989 increased surface runoff and contributed to a decrease in dry season flow. The rate of land degradation decreased and natural vegetation started recovering from the mid-1990s due to integrated watershed management interventions which resulted in an increased dry season flows and a declined surface runoff in the period 1989 to 2001. Whereas the wet season flows generated from surface runoff continued to decline in the most recent period (2001-2014), this was accompanied by an unexpected decline in dry season flow, which may be attributed to an increase in water withdrawals for irrigation. Analysis of 33 hydrological alteration parameters of simulated hydrographs from different LULC maps showed that the change in magnitude of median monthly flow, annual extremes, frequency and duration of flow pulses and rate and frequency of water conditions were consistent with the observed LULC changes over the period considered. In summary, the rate of increase in the peak flow and decrease in the dry season flow appeared to reduce after the 2000s. This result is attributed to the improvement of natural vegetation cover through watershed management interventions in the catchment.

The key finding from this study is that most LULC types are strongly affected the changes in hydrological components. Cultivation and bareland areas increase wet season flow and reduce dry season flow, AET and SWS. The reverse was found for natural vegetation cover (forest, wood and bush areas) which increases dry season flow, AET and SWS but decreases the wet season flow. The hydrological response to LULC change was more pronounced at sub-catchment level, which is mainly linked to the observed uneven spatial distribution of land degradation and rehabilitation in the catchment.

In conclusion, this paper has shown that ongoing watershed management interventions can increase dry season flows, while decreasing wet season flows. Dry season flows are of utmost importance for stakeholders as it comes when most needed. Stakeholders in the Geba catchment are already taking advantage of using some of the increased dry season flow for irrigation purposes. Further in-depth investigation of the impact of integrated watershed management intervention on the low flows is essential to understand the potential downstream implications, including in the Upper Tekeze basin and the Nile basin as a whole.

The approach applied in this study was found to be realistic to quantify hydrological responses to a human-induced environmental change in a complex catchment. Particularly, the development of a fully distributed hydrological model in wflow_PCRaster/Python modelling framework showed a good potential to simulate all hydrological components by maximizing available spatial data with little calibration to minimize the risks associated with over-parameterization. The PLSR model could subsequently identify how specific LULC types impact the different components of the hydrological cycle.

Chapter 7

MODELLING THE IMPACT OF CATCHMENT MANAGEMENT INTERVENTIONS ON THE LOW FLOWS [5]

The natural low flow of a catchment area is a valuable resource in semi-arid and arid lands. Large scale implementation of physical Soil and Water Conservation (SWC) structures can modify the hydrological processes of a catchment by changing the partitioning of the incoming rainfall on the land surface, both in positive and negative ways. The Geba catchment (5085 km²) in Ethiopia, forming the headwaters of Upper Tekeze basin, was known for its severe land degradation before the recent SWC interventions. These interventions have been successful as most of the degraded land is restored. On the other hand, the literature lacks studies which investigate the low flow response to SWC interventions and quantify the hydrological impacts at a larger scale. Understanding the response of low flows to SWC interventions is critical for effective water management policy interventions in the region. A combination of statistical test, Indicators of Hydrological Alteration (IHA) and catchment and model-to-model comparison approach was applied to understand the impact of SWC on low flows.

Results show that the catchment experienced a significant increase in the low flow following the ongoing intensive SWC implementation programs, whereas the low flow of the control catchment significantly decreased. Compared to the control catchment, low flow in the treated catchment was greater than that of control by more than 30% whilst the direct runoff was lower by more than 120%. This could be explained by the large proportion of the rainfall in the treated catchment that is infiltrated to recharge aquifers which subsequently contributes to streamflow during the dry season. The proportion of soil storage was more than double compared to the control catchment due to the SWC interventions. Hydrological comparison in a single catchment (model-to-model) also demonstrates that a drastic reduction in direct runoff (>84%) has improved the low flow proportion by more than 55% after the SWC works. However, despite the increase in low flow, the total streamflow has declined significantly after large scale SWC implementations which is attributed to the increase in evapotranspiration and soil storage.

This might have a negative impact on the availability of water resources to the downstream users. This entails that implementation of catchment management strategies should consider to enhance the availability of water resources in both upstream and downstream catchments.

---

[5] Based on: Gebremicael, T.G., Y.A. Mohamed, P. van der Zaag, submitted. Modifications of low flow due to catchment management dynamics in the headwaters of Upper Tekeze basin of Ethiopia. A comparative catchment modelling approach (Submitted to *Hydrological Processes*)

7.1 INTRODUCTION

Low flows are the actual flow in a river during the prolonged dry period or during periods without rainfall (Bradford & Heinonen, 2008; Smakhtin, 2001). Low flows are derived from groundwater and from surface runoff from lakes and marshes; whereby groundwater flows is often the major contributor to the total dry season flow in a catchment (Wittenberg, 2003; Douglas et al., 2001). The magnitude and variance of low flows depend on the seasonal distribution of rainfall as well as inter-seasonal variability (Giuntoli et al., 2013; Pushpalatha et al., 2012). Accurate estimates of low flow characteristics in a catchment is fundamental for water resources management (Castiglioni et al., 2011; Laaha & Blöschl, 2006). For example, feasibility for micro-hydroelectric plants, environmental flow requirements, irrigation water withdrawals, domestic water use rely on availability and consistency of low flow. This is why much of the focus on water management, in particular in arid and semi-arid region, has been on finding the balance between the incoming and outgoing water from rivers during the low flow periods (Giuntoli et al., 2013; Smakhtin, 2001).

Low flow varies in response to catchment natural controls on runoff (e.g., rainfall, geology, topography, soil) and anthropogenic disturbances (Guzha et al., 2018; Gebremicael et al., 2013; Giuntoli et al., 2013). A combination of human activities that include land use/cover (LULC) change, direct water abstraction from surface water and groundwater, hydropower operation rules, soil conservation interventions and water harvesting systems can modify the overall streamflow and the low flow in particular (Chang et al., 2016; Li et al., 2007). Water withdrawals for industry and irrigation during the dry season significantly reduce the availability of low flows in a river whereas change in land use/cover can both increase and decrease the low flows (Gashaw et al., 2018; Guzha et al., 2018). Conversion of natural vegetation cover into agricultural and grazing land can cause a decrease of low flows, but increase direct runoff (Woldesenbet et al., 2017; Gebremicael et al., 2013; Bewket & Sterk, 2005).

Soil and Water conservation (SWC) interventions, such as building terraces, sediment trapping dams, trenches and changing land cover by replanting trees and improving pastures, cause visible changes in the stream flow regime (Wang et al., 2013; Li et al., 2007; Mu et al., 2007). Large scale implementations of such interventions can modify the hydrological processes of a catchment by changing the partitioning of the incoming rainfall at the land surface (Gates et al., 2011). Introduction of physical SWC structures (terraces, bunds, trenches etc.) can reduce the surface runoff by up to 86% whereas it can increase the base flow by more than 50% (Abouabdillah et al., 2014; Schmidt & Zemadim, 2013; Lacombe et al., 2008; Mu et al., 2007).

However, there is no distinct understanding in the literature on how biophysical SWC intervention (vegetation) affects the dry season flow. Wang et al. (2013), Hengsdijk et al. (2005) and Huang & Zhang, (2004) reported that, improving catchments with plantation and pastures can increase the dry season flow by enhancing infiltration capacity during the rainy season whilst other researchers showed a decreasing of low flow due to increase in interception and actual evapotranspiration (e.g., Silveira & Alonso, 2009; Brown *et al.*, 2005; Huang & Zhang, 2004). In semi-arid catchments, an increase in low flow is reported after improving vegetation cover (Nyssen *et al.*, 2010; Huang & Zhang, 2004). It can improve green water use efficiency and groundwater recharge at a local level while reducing total surface runoff at a larger scale (Garg *et al.*, 2012). Conversely, it may enhance subsurface flow which increases dry season flow at the larger scale. Such conflicting results call for further investigation on the impact of SWC interventions on low flows at different spatial scales is vital for improved water management.

The Upper Tekeze headwater catchments are characterized not only by severe land degradation and moisture stress but also known for the recent integrated catchment management experience (Gebremeskel *et al.*, 2018; Belay *et al.*, 2014; Chapter 5). Since the mid-1970s, various land and water management interventions have been implemented to enhance food security and environmental rehabilitation in the basin (Gebremeskel *et al.*, 2018; Woldearegay *et al.*, 2018; Asfaha *et al.*, 2014). However, the majority of the SWC programs implemented before the mid-1990s were little successful due to technical and institutional problems (Woldearegay *et al.*, 2018; Alemayehu *et al.*, 2009). The top-down approach of SWC implementation programs during this period contributed to a limited adoption and acceptance by the local community (Woldearegay *et al.*, 2018; Bishaw, 2001). Since the mid-1990s, integrated catchment management approaches that include physical and biological SWC interventions were introduced (Nyssen *et al.*, 2015a; 2014; Asfaha *et al.*, 2014; Alemayehu *et al.*, 2009). These interventions resulted in the restoration of extensive areas with severe land degradation (Gebremeskel *et al.*, 2018; Guyassa *et al.*, 2018; Woldearegay *et al.*, 2017; Nyssen *et al.*, 2010; Alemayehu *et al.*, 2009;).

Integrated catchment management interventions in upstream catchments have increased infiltration of rainwater and the discharge of springs and streams in lower parts of catchments (Gebremeskel *et al.*, 2018; Fanta *et al.*, 2017; Haregeweyn *et al.*, 2015; Nyssen *et al.*, 2010; Descheemaeker *et al.*, 2006b). Studies from runoff experimental plots (2m^2 - 2km^2) at different locations in the Geba catchment showed that the availability of water during the dry season has significantly increased after the implementation of these interventions (Taye *et al.*, 2011; Nyssen *et al.*, 2010; Descheemaeker *et al.*, 2006b;). Such changes were observed in terms of increasing groundwater levels, decrease of direct surface runoff, emerging of springs and expansion of household irrigation from shallow groundwater sources. Furthermore, field observations by Alemayehu *et al.* (2009) at upper Agula catchment (14.5km^2) indicates that surface runoff has significantly reduced while groundwater has significantly improved after SWC implementations. Fanta *et al.* (2017) found a decreasing and increasing trend of annual and dry season flows, respectively, in Agula catchment between the period of 1992 and 2012. The authors showed that the drivers of these changes were due to human interventions in the catchment. These achievements can also be evidenced by the expansion of small-scale irrigation schemes in the basin (Gebremeskel *et al.*, 2018; Haregeweyn *et al.*, 2015; Nyssen *et al.*, 2010; Alemayehu *et al.*, 2009).

However, the literature showed limited studies to quantify the impacts at larger scale or catchment level. Results from experimental plots, surveys or micro-watershed levels ($<2km^2$) may not be extrapolated to basin scale (Lacombe *et al.*, 2008). In this respect, the literature lacks studies that investigate the low flow response to SWC interventions. As the impact of SWC interventions is more pronounced on the base flow, improved scientific understanding on the response of dry season flow to SWC interventions is critical for effective water management policy interventions in the basin.

The magnitude of change in streamflow due to change in various land management practices can be quantified using time series trend analysis, hydrological models, multivariate statistics and paired catchments (Gashaw *et al.*, 2018; Saraiva Okello *et al.*, 2015; Kiptala *et al.*, 2014; Shi *et al.*, 2013). A paired catchment analysis has been commonly applied to identify the difference in hydrological responses to catchment disturbances due to SWC interventions, water harvesting structures and change in vegetation covers (Worqlul *et al.*, 2018; Zhao *et al.*, 2010). The basic concept of paired catchment involves the comparison of hydrological response of two adjacent catchments (one as a control and other as a treatment); alternatively, the hydrological response "before and after" interventions in a single catchment can be compared (Ssegane *et al.*, 2013; Bishop *et al.*, 2005; Brown *et al.*, 2005). In a paired catchment approach, it is not necessary that the two catchments are identical, but are comparable in characteristics such as topography, area, soil, climate, vegetation cover and in close proximity to each other (Zhao *et al.*, 2010; Best *et al.*, 2003). Comparison of hydrological response from paired catchments is not always reliable to determine the absolute value of streamflow, but it is the foundation of many studies to ascertain the differences in hydrological responses between catchments having different land management interventions (Kralovec *et al.*, 2016; Zhao *et al.*, 2010; Brown *et al.*, 2005; Watson *et al.*, 2001).

The main goal of this study is to investigate low flow responses to integrated catchment management interventions in the Geba catchment of Upper Tekeze headwaters using a paired catchment comparison approach. A combination of comparisons from two adjacent catchments (one with intensive SWC interventions and the other as a control) and single catchment ("before and after" interventions) was used to understand the impact of interventions on the dry season flow of the catchments. Furthermore, spatial and temporal variations of streamflow, and in particular of the dry season flow, were examined using a spatially distributed rainfall-runoff model. The results from these two catchments were confirmed by a model-to-model approach that compares simulated hydrological patterns with parameter sets calibrated for the period before and after SWC interventions in a single catchment.

7.2 STUDY AREA DESCRIPTION

This study was conducted in two comparative catchments; Agula (481 km²) and Genfel (502 km²) within the Geba catchment in Ethiopia (Figure 7.1). The outlets of the two adjacent catchments are close to each other at a distance of less than 5km. They are located in northern

Ethiopia between 13.54°N, 39.59°E to 14.14°N, 39.80°E in the headwaters of Upper Tekeze, a tributary of the main Nile river basin (Figure 7.1). Table 7.1 summarizes the characteristics of the paired catchments. The topographic characteristics such as elevation, size, slope, length of stream network and drainage density of the two catchments are very close to each other. They both have similar features of physical geography and hydrography in the altitude between 1961 and 3070 m.a.s.l. (Figure 7.1). They are also very similar in terms of geological information in which Enticho Sandstone, Adigrat Sandstone, Antalo Super sequence and Meta-morphic (basement) rocks are dominant in both catchments (Gebreyohannes et al., 2013). However, the major difference is depicted by average annual discharge and percentage coverage of SWC measures.

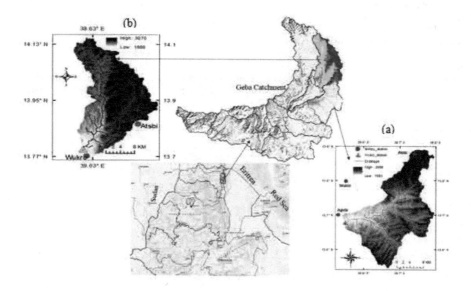

Figure 7.1: Location of the study area, meteorological and hydrological monitoring stations, (a) Agula, (b) Genfel

The two catchments are characterized by a semi-arid climate in which the majority of the rainfall occurs from June to September after a long dry season (Chapter 4). More than 70% of the total rainfall falls in July and August only, mainly during high- intensity storm events. Rainfall over the two catchments is highly variable mainly associated with the seasonal migration of the intertropical convergence zone (ITCZ) and the complex topography (Nyssen et al., 2005). However, compared to each other, they have almost the same mean annual rainfall and temperature. The land use/cover of the two paired catchments are also comparable. A slightly higher vegetation cover (bushes and shrubs) and less bareland in Agula is due to the intensive watershed management interventions in the catchment (Chapter 5). The same agricultural systems are practised in the two catchments wherein farmers use a mixed subsistence farming based on crops and livestock production. Although the landholding size of each crop can be different, the same type of crops are grown in the two catchments.

Table 7.1: Catchment characteristics of the two catchments after management interventions

Catchment characteristics	Agula	Genfel
Catchment area (km²)	481	502
Maximum elevation (m.a.s.l.)	2900	3070
Minimum elevation (m.a.s.l.)	1961	1988
Average slope (%)	17.3	19.8
Area with slope >30% (km²)	16.5	14.9
Drainage density (km/km²)	1.93	2.05
Agro-climatic zone	Semi-arid	Semi-arid
Length of major river (km)	41	46
Mean annual temperature	21.6	21.3
Annual average discharge (m³/s)	1.1	0.6
Average dry season flow (m³/s)	0.6	0.2
Mean annual rainfall (mm/year)	550	560
Potential evapotranspiration (mm/year)	1430	1432
Aridity index (-)	0.39	0.39
Moisture index (%)	-59	-59
Major soil types	Leptosol, Cambisol, Luvisol	Leptosol, Cambisol, Luvisol
Major crops	teff, wheat, barley, maize, tomato, potato	teff, wheat, barley, maize, tomato, potato
Proportion of land use/cover types (%)		
Rain fed agriculture	30	34.2
Irrigated agriculture (from 2016)	1.4	0.8
Forest land	2.5	1.2
Bushes and shrubs	26.4	19
Wood land	11.7	8.4
Plantations (eucalyptus trees)	5.9	7.5
Bare land	8.4	17.6
Grass land	13	10.9
Urban areas	0.6	0.3
Water bodies	0.2	0.1
SWC Coverage (%)	49.6	15.4

The soil is one of the critical features that control the hydrologic response of a catchment to the incoming rainfall events (Worqlul et al., 2018). Cambisol, Luvisol and Leptosol are the dominant soil type in both catchments i.e. 33%, 25% and 25% in Agula and 36%, 30% and 18% in Genfel, respectively. In general, weathered soils are found in the uppermost plateaus, rocky and shallow soils in the vertical scarps, coarse and stony soils in the steep slopes, finer textured soils in the undulating pediments and most deep alluvial soils are found in the alluvial terraces and lower parts of the alluvial deposits (Gebreyohannes et al., 2013).

In summary, whereas both catchments are very similar, they differ in the proportion of land management interventions that was subject to catchment rehabilitation interventions and annual discharge. The level of SWC interventions between the two catchment is significantly different (Table 7.1). The total SWC physical structures in Agula is 49.6% while it is only 15.4% in

Genfel. This allows for a paired catchment comparison since hydrological response can differ due to difference in these land management interventions.

Over the last two decades, the government and NGOs have given strong attention to rehabilitate the degraded land with integrated watershed management interventions in the Geba catchment (Gebremeskel *et al.*, 2018; Woldearegay *et al.*, 2018; Belay et al., 2014). After the mid-2000s, an integrated approach of SWC interventions including, terraces, water harvesting structures, check dams, gully rehabilitations, deep trenches, percolation ponds, and revegetation through area enclosures have been introduced at different locations in the catchment (Figure 7.2).

Figure 7.2: Physical SWC interventions in the Geba catchment of the Upper Tekeze headwaters: (a) terraces, (b) check dams, (c) water harvesting structures, (d) percolation ponds at the top of hills, (e) gully rehabilitation using gabion, (e) deep trenches at the bottom of hills (sources: (Woldearegay K & Tamene L, 2017; Yosef & Asmamaw, 2015).

These interventions have been implemented by mobilizing the community for free-labour days and food for work through different programs including; Managing Environmental Resources to Enable Transition to more sustainable livelihoods (MERET), Productive Safety Net Program (PSNP), and the National Sustainable Land Management Project (SLMP) (Haregeweyn *et al.*, 2015). These interventions may change the rainfall-runoff relationships of the catchment. Change in the water–table level and altering the partitioning of rainfall between surface and groundwater may occur at a various time and spatial scales depending on the local context. In fact, the Agula catchment has received SWC interventions at a significantly larger scale than Genfel which allows for a paired catchment comparison. Given the large differences in the level of intervention between the paired catchment, a paired catchment comparison is possible as the low flow response and water availability could be different in the two sub-catchments.

7.3 DATA AND METHODS

The low flow responses to integrated catchment management interventions in the Geba catchment of Upper Tekeze headwaters were analysed using different approaches. To understand how the interventions changed the low flows in the catchment, first, the relationships of the observed flows from before and after the interventions were quantified using different Indicators of Hydrological Alteration (Mathews and Richter, 2007) parameters, and Pettitt (Pettitt, 1979) and Mann–Kendall (MK; Kendall, 1975) statistical tests. However, these methods do not show how the SWC interventions influence the overall hydrological processes of the catchments. To infer the physical mechanisms behind the changes, if any, a catchment modelling comparison approach was applied. A spatially distributed hydrological model based on Wflow-PCRaster/python framework were developed to simulate and compared the overall hydrological responses and in particular the low flow from the two catchments (the detailed description and development of the model is presented in Chapter 6). Static data that contain maps that do not change over time and dynamic data containing maps that change over time are the main inputs for PCRaster/python framework based Wflow distributed hydrological model development. These datasets are explained in detail in the following sections.

7.3.1 Static dataset

Land use/cover, Digital Elevation Model (DEM), soil, Local Drainage Direction (LDD) and river maps are the main static input datasets required to develop a distributed hydrological model in a Wflow_PCRaster/Python framework environment (Schellekens, 2014, Hassaballah et al., 2017; Chapter 6).

The land use/cover maps of 2003 and 2014 were developed for both sub-catchments (Figure 7.3). These years were selected considering the intensive watershed management interventions in the catchments from around 2005/2006 (Woldearegay et al., 2018). Landsat7 and 8 images for 12/7/03/2003 and 23/03/2014, respectively, were obtained from the US Geological Survey (USGS) center for Earth Resources Observation and Science (EROS) found in http://glovis.usgs.gov/. Images from the same period were selected in order to minimize the seasonal effect during image pre-processing and classifications. In addition to the high-resolution satellite images, ancillary data including field observations, topographic maps and secondary literature were collected from different sources. Ground truth points (200 each) used for the classifications and evaluation of the accuracy of classified satellite images were collected from our previous study (Chapter 5) and a field survey during 15/09/2018 to 10/11/2018. Before classification, image pre-processing such as geometric and atmospheric corrections were done in ERDAS Imagine software. The same procedure as described in Gebremicael et al. (2018) was applied to identify the different land use types shown in Figure 7.3. Both supervised and unsupervised classification approaches were applied to classify the images. An Iterative Self-Organizing Data Analysis Technique (ISODATA) clustering algorithm was selected to identify the spectral classes during unsupervised classification (Wafi,

2013; Cheema & Bastiaanssen, 2010). The classified LULC maps from unsupervised classifications were further refined with expert judgment and field observations during the supervised classifications.

Finally, the Maximum Likelihood algorithm was used for the classification of all images in the supervised classifications (Al-Ahmadi & Hames, 2009). The final classified LULC maps were evaluated using independent ground truth data randomly collected from the fieldwork and our previous study (Chapter 5). The final produced LULC maps (Figure 7.3) were used as an input for the development of a distributed hydrological model.

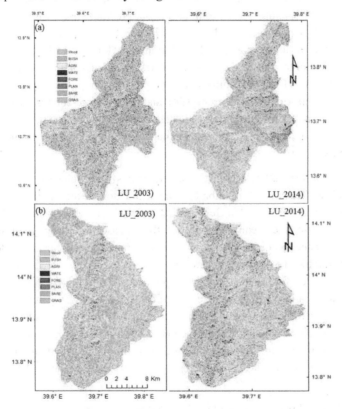

Figure 7.3: Land use/cover maps used for model inputs, (a) Agula catchment, (b) Genfel Catchment

Soil maps of the paired catchments were obtained from our previous study (Chapter 5). The initial soil map was obtained from the International Soil Reference and Information Centre (ISRIC) SoilsGrid250 (Hengl *et al.*, 2017) and modified based on the study area characteristics. A detailed description can be found in Gebremicael *et al.* (2018). All the static input maps were projected to WGS-84-UTM-zone-37N and resampled to a resolution of 50m for the model inputs. Catchment characteristics format needed for the Wflow model were created from the

DEM, land cover, soil and hydrological gauge locations with a pre-preparation step1 and step2 of the WFlow model (Chapter 5).

7.3.2 Dynamic datasets

Daily precipitation (P) and potential evapotranspiration (PET) are the main dynamic input data to force the hydrological model. Open access satellite-based rainfall and evapotranspiration products were used in this study as there is not enough ground observed climatic data in both sub-catchments. The CHIRPS satellite rainfall product was used in this study. A comparison of eight freely available satellite rainfall estimates in our previous study (Chapter 3) showed that the CHIRPS product performed well over the Upper Tekeze basin. The CHIRPS rainfall estimates and PET are found at a spatial resolution of 5.5km and 11 km, respectively. For a detailed description of the CHIRPS rainfall estimates, refer to Gebremicael et al. (2019b). The PET data were obtained from the Famine Early Warning System Network (FEWS NET). These data are available on a daily basis from 1981 and 2001 to present for CHIRPS and PET, respectively. The satellite-based rainfall estimates and PET maps were projected to WGS-84-UTM-zone-37N, in the same way as the static maps, and resampled to a resolution of 50m for the model inputs.

Streamflow data at the outlets of the paired catchments were collected from the Ethiopian Ministry of Water Resources for calibration and validation of the model. Descriptions and quality of these data are presented in Gebremicael et al. (2017). Since around 2010/11, irrigated agriculture in the two catchments has significantly increased and hence the low flow in the rivers has been reduced due to abstraction for irrigation consumption (Chapter 6), it was necessary to naturalize the observed streamflow. In this study, abstracted water during the dry months was estimated from actual evapotranspiration (AET) derived from remote sensing. Actual evapotranspiration from irrigated pixels of the study areas was generated using Landsat information and PySEBAL model (Jaafar & Ahmed, 2019; UNESCO-IHE, 2018). The foundation of PySEBAL model is the Surface Energy Balance Algorithm for Land (SEBAL) which requires spatial information in the visible, near-infrared and thermal infrared as well as spatially distributed weather data (Van Eekelen et al., 2015; Kiptala et al., 2013;Bastiaanssen et al., 2005;).

The PySEBAL model was developed based on Python open source platform and it is capable of semi-automatic processing of selected satellite images. The advantage of PySEBAL over the original SEBAL is the improvement to include a function on an automatic selection of hot and cold pixels as well as inclusiveness of biomass and crop estimates. The PySEBAL translates raw satellite measurements into maps of actual evapotranspiration for every 30 x 30 m spatial resolutions. Previous research has shown that SEBAL has great potential to estimate AET from irrigated agriculture (Eekelen et al., 2015; Kiptala et al., 2013). The detailed information on the algorithm embedded in PySEBAL is found in Bastiaanssen et al. (2005; 1998a). The time step of 16 days with 30m spatial resolution of Landsat images was used to estimate the AET fluxes during irrigation seasons from 2011 to 2016. This time step is limited by the availability of

Landsat products. Considering the local irrigation practices, it is a reasonable time scale and the spatial resolution corresponds to the small scale irrigation practices in the catchments.

The final estimated AET at 16 days interval were interpolated into daily time step using linear interpolation and converted to daily discharge at the paired catchment outlets. Considering the local irrigation practices, 70% of application efficiency was taken to convert the actual evapotranspiration into discharge. An average daily discharge of 0.16 and 0.13 m^3/s were obtained from the irrigated pixels of Agula and Genfel during the irrigation period, respectively. The naturalized daily discharge at the catchment outlets were obtained by adding the observed and the estimated ET from the irrigated agriculture and this data were used for the calibration and validation of the model. The performance of the model has improved when calibrated using the naturalized streamflow. For example, the Nash-Sutcliffe efficiency of the model in Agula catchment was improved from 0.81 to 0.92 after naturalization of the streamflow from the irrigated pixels. An example of comparison of the original and naturalized streamflow during the irrigation period is given as supplementary file in Appendix E (Figure E-1).

7.3.3 Methods

Model calibration and validation

The developed hydrological model for each catchment was calibrated and validated using independent daily input data at the outlets (Figure 7.1). The calibration parameters were selected based on our previous work in Chapter 6 and recommendations from other literature (Hassaballah et al., 2017; Schellekens, 2014). Initial parameter values were taken from Chapter 6, own field observation and laboratory analysis. Next, parameter values were manually adjusted until maximum concordance between observed and predicted streamflow occurred. Climatic and hydrologic datasets of 2001 to 2006 and 2011 to 2016 were used for calibration and validation of models during the before and post interventions comparisons, respectively. The first three years of each period were used for calibration while the remaining three years were used for validation of the models. The different models were then evaluated using different objective functions to verify whether the predicted and observed streamflow agreed. Nash–Sutcliffe Efficiency (NSE) and Percent Bias (PBIAS) statistical indices were applied to evaluate the performance of the model. Detailed descriptions of these indices are given in our previous study (Chapter 3).

Analysis and comparison approach

First, Pettitt and MK tests and IHA method were applied to understand if there is a significant change in the naturalized streamflow of the paired catchments by considering before and after the interventions. The statistical test was used to identify if there is an abrupt change of streamflow after the interventions whilst the IHA method was used to assess alteration of low flow in both catchments. To further understand the physical causes of these changes (if any), the calibrated and validated distributed hydrological models were applied to simulate the low flow response to the physical SWC interventions in the paired catchments. The response of low flows were studied in different steps: (a) Paired catchment comparison using two independent catchments, (b) Paired catchment comparison using pre and post-catchment interventions, (c) Evaluation of calibrated model parameters if the differences in catchment management interventions were reflected in the low flow responses.

The relative rate of low flow response to catchment management between the two independent catchments was compared. Despite similarities of catchment characteristics, the rate of low flow response is expected to vary between the two catchments. Next, water balance models focusing on the low flow from before and after interventions within each catchment were compared. The difference of the water balance components from both periods was analysed. Furthermore, model parameters between before and after interventions in each catchment were also investigated to assess whether the values changed due to SWC interventions or not.

7.4 RESULT AND DISCUSSION

7.4.1 Comparison of change in streamflow

The Pettitt and MK tests ($P < 0.05$) and IHA method were applied to the observed annual, wet and dry season streamflow of the period 1996 to 2016 for the two catchments. The pattern and comparison of seasonal and annual streamflow in the paired catchments are given in Figure 7.4 and Table 7.2. The annual and wet season (June-September) streamflow of the two catchments significantly reduced for the given period (Figure 7.4). The decline in annual and wet season flows during this period could be attributed to an increase in proportion of evapotranspiration and ground water recharge which subsequently contributes to streamflow during the dry seasons. The changes in streamflow without significant change in precipitation, as shown in our previous study (Chapter 4) indicates factors other than precipitation are the main drivers of the change in streamflow. The dry season flow of Genfel has significantly decreased; by contrast, the dry season flow of Agula has significantly increased for the same period of analysis.

The result from the MK test is also consistent with the Pettitt test, namely that both annual and wet season flow showed a decreasing trend in both catchment (Table 7.2). Positive (negative) values of Z statistics associated with the computed probability (P-value) indicate an increasing (decreasing) trends of the flow. The result is in agreement with our previous work in Geba catchment and with Fanta et al. (2017) in Agula catchment who reported a change in streamflow without significant change of rainfall. The different result in the dry season of the paired catchment suggests that the enhanced dry season flow in Agula catchment could be attributed to the modifications of catchment responses through SWC practices.

Table 7.2: Summary results of MK test on streamflow trends. Negative (positive) Z value indicates a decreasing (increasing) trend and statistically significant trends at 5% confidence level are shown in bold (Z=±1.96).

Station name	Annual flow	Wet season flow	Dry season flow
Agula	**-4.4**	**-3.4**	**3.2**
Genfel	**-3.8**	**-2.4**	**-2.1**

The effect of SWC interventions on the hydrological regimes of the study area was quantified by IHA method, using the naturalized observed streamflow at the outlet. To understand how the interventions modified the low flows in the catchment, the relationships between the flows before and after the interventions were quantified using different IHA parameters (Figure 7.5). In agreement with the statistical test, the result from IHA ascertained that there was a significant change in hydrological variables after the SWC works in the Agula catchment. The median monthly flow during the dry seasons increased up to 78% (Figure 7.5a). In contrast, the median monthly flow during the peak season declined by more than 50% after the intensive SWC interventions.

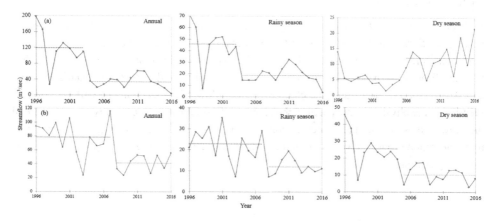

Figure 7.4: Homogeneity test of annual and seasonal naturalized streamflow in the paired catchment, (a) Agula, (b) Genfel

Similarly, all indicators for low flow of annual conditions exhibited a significant change in magnitude and duration between the two periods. The summary of annual flow conditions for Agula catchment is presented in Figure 7.5 whereas for the Genfel catchment, it is given as supplementary file in Appendix E (Figure E-2). The values of all except the 90-day annual minima parameters were negligible before interventions whilst increased up to 0.6 m^3/s after the interventions (Figure 7.5b). Similarly, all maxima parameters moderately declined in the post-treatment period (Figure 7.5c). The decrease in annual maxima and increase of annual minima parameters between the two periods shows the influence of catchment management interventions in the catchment. The base flow index i.e., the ratio of 7-day minimum flow to the mean flow of the year increased from 0 to 0.11 m^3/sec between the two periods. This indicates that the contribution of subsurface flow to the stream has increased after SWC development in Agula catchment.

Assessment of the alteration of frequency and duration of low pulses indicates that their magnitude and duration were higher in the pre-treatment compared to the post-treatment (Figure 7.5d). This shows the low flow in the Agula catchment is becoming constant after the implementation of SWC measures. These, in turn, reflect that the rainfall contribution to the groundwater increased in the catchment. In contrast, the high pulse count above and below the given threshold during the pre-treatment period demonstrates the hydrological response of the catchment was flashier. The number of high and low flow pulses is associated with the rise and fall rates and gives evidence on how the low flow response is increasing after the physical SWC interventions. The median rise rate (positive differences between two consecutive daily values) and fall rate (negative difference) has decreased from 0.15 and 0.25 m^3/sec to 0.08 and 0.11 m^3/sec, respectively. This denotes that the frequent fluctuation of low flows during the pre-treatment period has decreased and more stable flow is contributed from the catchment after the interventions.

In summary, all indicators showed a very high (>60%) alteration in low flow between the two periods which reveals the impact of SWC practices implemented in the period between the mid-2000s and mid-2010s. The overall change in the hydrological alteration parameters elucidate the modification of the flow regimes of the catchment. The findings from the statistical tests and IHA method demonstrated the change in streamflow of the catchments, without however identifying the physical causes behind the changes. To establish such causality, the response of the low flows to human interventions such as large-scale catchment management implementation programs were analysed using a paired catchment modelling approaches in the following sections.

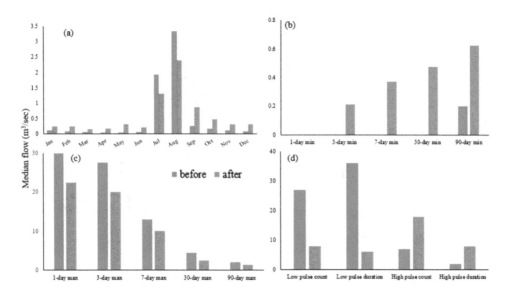

Figure 7.5: Comparison of hydrological responses in Agula catchment before and after catchment management interventions using indicators of hydrological alteration parameters: (a) Magnitude of average median monthly flow; (b) Magnitude of median annual average minimum flows, (c) Magnitude of median average annual maximum flows and (d) Frequency and duration of average high and low pulses

7.4.2 Model calibration and validation

Daily streamflow hydrographs during the two calibration (2001 to 2003 and 2011 to 2016) and validation periods (2004 to 2006 and 2014 to 2016) are given in Figure 7.6 and 7.7, respectively. Results presented in these figures indicate generally a good agreement between the simulated and observed streamflow in the paired catchments. The models of both catchments were able to capture the daily hydrographs both for the peak and low flows. The simulated and observed hydrographs have very similar variation trends in Agula and Genfel catchments in both simulation periods. However, the consistent slight overestimation of the flow across all simulation years in both catchments. The consistent overestimation of streamflow could be related to the coarse resolution of evapotranspiration from the satellite products or due to model structure errors.

The statistical indicators for the evaluation of model performance are presented in Table 7.3, showing a satisfactory model performance. The NSE value at both catchments are greater than 0.8 whilst values of PBIAS and RMSE are below 20% and 10, respectively. Model performance in streamflow prediction is acceptable as a satisfactory result if NSE > 0.5 and PBIAS value within ±25% (Moriasi et al., 2007). This indicates our model performance is excellent both for the calibration and validation periods during 2001-2006 and 2011-2016 simulation periods.

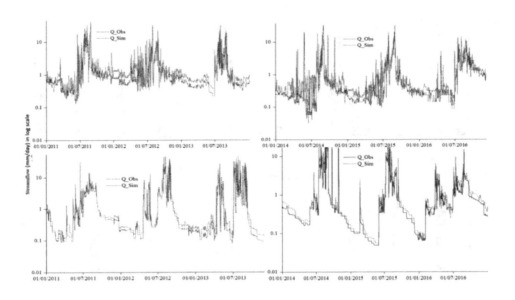

Figure 7.6: Calibration (right) and validation (left) of the wflow-sbm hydrological model at Agula (a) and Genfel (b) during the 2011-2016 period

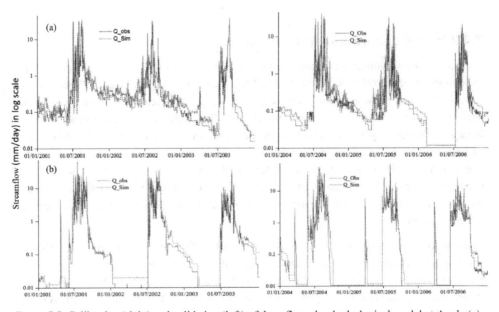

Figure 7.7: Calibration (right) and validation (left) of the wflow-sbm hydrological model at Agula (a) and Genfel (b) during the 2001-2006 period

Table 7.3: Evaluation performance of the hydrological model during calibration and validation

Simulation period		Catchments					
		Agula			Genfel		
		NSE	PBIAS	RMSE	NSE	PBIAS	RMSE
2011-2016	Calibration	0.92	8	2.3	0.89	7	4.2
	Validation	0.89	11	5.8	0.87	14	6.8
2001-2006	Calibration	0.87	12	5.1	0.84	19	7
	Validation	0.85	17	6.3	0.82	21	10

The performance of the model seems to be better than from our previous study in the same study area (Chapter 6). This may be explained that the observed streamflow in this study was naturalized using remote sensing information whereas observed streamflow in the previous study was affected by water withdrawals for irrigation during the dry season which can affect the performance of the model. Overall, the good agreement between simulated and observed streamflow indicates that the wflow-sbm distributed hydrological model is applicable for the analysis of hydrological processes in the paired catchments. While the most sensitive model parameters are discussed in our previous study (Chapter 6), the values used in this study are shown in Table E-1.

7.4.3 Comparison of water budget of the paired catchments

The simulated average water balance and proportion of each hydrological component of Agula and Genfel catchments are presented in Table 7.4. The average absolute value differences between the water balance components of Agula (treated) and Genfel (control) is provided for the comparison. On average, both catchments received the same amount of precipitation and they evaporate similar proportional amounts of water. However, Genfel catchment exhibited higher runoff volume (32%) than Agula catchment (13%), while the reverse is true for the base flow (the base flow index BFI is here used as a proxy) and soil water storage. The annual average streamflow of Agula was greater than that of Genfel, as shown Table 7.1, but this has changed after catchment management interventions when the annual flow of Agula significantly declined compared to Genfel (Table 7.4). Despite the similarities in climatic characteristics, the ratio of runoff volume in Genfel is higher than Agula by more than 80%, while the amount of precipitation contributed to base flow is lower by more than 60%.

The ratio of base flow to the total discharge in Agula is almost double that of Genfel, suggesting more of the incoming rainfall is contributing to the groundwater recharge. The overall runoff coefficient in Agula is lower than in Genfel by 80% which implies more of the incoming rainfall is contributed into soil water storage which is on average greater than from Genfel by more than 90%. This higher soil storage comes from the drastic reduction of runoff response to rainfall in Agula. The proportion of soil storage to the total incoming rainfall in Agula is more than double compared to Genfel catchment. The result is very important because the amount of incoming rainfall to both catchments are almost the same (Table 7.4) with similar seasonal variations. Therefore, the observed differences in the hydrological components must be related to the catchment management interventions and overall storage properties in the catchment.

The hydrological processes in the two catchments were further analysed by looking into seasonal hydrological variability. The proportion of runoff fluctuated depending on the amount of rainfall and seasons. The greater differences in runoff proportion between the two catchments were in wet season (June-September) when more than 80% of the annual rainfall occurred. In contrast, the lowest values and smallest differences were found during the dry season (October – May). The runoff proportion in Agula is lower than in Genfel catchment by more than 120% during the rainy season which suggests more of the input rainfall in Genfel is going to runoff production compared to Agula catchment. On the other hand, large parts of the seasonal rainfall in Agula is infiltrated into groundwater which later contributes to streamflow during the dry seasons. This is also ascertained by the large differences in water storage (38%) between the two catchments during the driest months (January-May). The large differences in water storage in the dry season is attributed to the surface characteristics that enhance the infiltration capacity of the catchment. Moreover, a noticeable difference in streamflow is also observed during the fall months (October- December) i.e., the recession flow in Agula is higher than in Genfel by more than 20%.

In summary, higher percentage differences during the dry season indicates the contribution of groundwater flow to Agula is greater than that of Genfel. This implies that the streamflow in Agula is distributed more homogeneously among the different seasons compared to Genfel which more than 80% of the total flow comes from the rainy season (July-September) only (Table 7.4).

Table 7.4: Average simulated water balance components of the paired catchment during 2014-2016 simulations and their difference (after catchment management interventions).

Water budget components	Genfel catchment		Agula catchment		Difference
	Average	%	Average	%	
Annual precipitation (mm/year)	634	100	642	100	8
Actual evapotranspiration (mm/year)	355	56	385	60	30
Deep percolation (mm/year)	43	7	116	18	73
Change in storage (mm/year)	33	5	57	9	24
Annual total flow (mm/year)	203	32	84	13	-119
dry season flow (mm/season)	27	4	37	6	10
Wet season flow (mm/season)	176	28	47	7	-129
Runoff coefficient (-)	0.32		0.13		-0.19
Base flow index (BFI)	0.28		0.53		0.25

The relationship between the hydrological components before catchment management interventions were also compared for the paired catchments (Table 7.5). The results indicate that the remarkable differences in the hydrological response of the two catchments during the period from 2014-2016 were not visible before the intervention (2004-2006) programs. With almost the same precipitation inputs (≈ 1% difference), surface runoff, evapotranspiration and base flow responses did not show substantial differences between the two catchments. An interesting result is that unlike after the intervention programs, the base flow index of Genfel was greater than Agula by 18%. This implies that the two adjacent catchments had the same hydrological response during the 2004-2006 period. This proves that the observed large differences in the hydrological fluxes during the 2014-2016 simulation period could be attributed to the large scale physical SWC structure implementation programs which have significantly influenced the partitioning of incoming precipitation and soil storage.

Table 7.5: Average simulated water balance components of the paired catchment before catchment management interventions (2004-2006 simulation period).

Water budget components	Genfel catchment		Agula catchment		Absolute difference
	Average	%age	Average	%age	
Annual precipitation (mm/year)	536	100	540	100	4
Actual evapotranspiration (mm/year)	214	40	222	41	8
Deep percolation (mm/year)	38	7	43	8	5
Change in storage (mm/year)	-14	-3	-9	-2	5
Annual total flow (mm/year)	298	56	284	53	-14
dry season flow (mm/season)	36	12	22	8	-14
wet season flow (mm/season)	262	88	262	92	0
Runoff coefficient	0.56		0.53		-0.03
Base flow index (BFI)	0.18		0.15		-0.03

7.4.4 Model to model comparison approach

The rainfall-runoff process relationship before and after the treatment programs of the paired catchments was analysed using a model-to-model (pre- and post-treatment) comparison approach. This model-to-model comparison approach compares the hydrological response of the catchments with parameter sets calibrated before and after management interventions. Results shown in Table 7.6 indicate that Agula catchments experienced an increase in low flows and soil storage whilst a decrease in surface runoff following environmental rehabilitation programs. Unlike in Agula, the low flow in Genfel catchment showed a decreasing pattern after the intervention. It is explicitly shown in the analysis that the hydrological behaviour of Agula catchment changed more dramatically compared to Genfel catchment. This may be explained by the much more extensive physical SWC interventions in the catchment (Section 7.2).

A total reduction in naturalized streamflow by 70% is observed between the pre- and post-treatment periods (Table 7.6). The significant reduction in total streamflow is due to the increase in actual evapotranspiration (73%) and significant soil storage enhancements after physical and biological SWC interventions. Surface runoff contribution to the river discharge of Agula has significantly reduced (82 %) between the two model validation periods (2004-2006 and 2014-2016). In contrast, low flows during the dry season have increased up to 68% after the interventions. Improvement in catchment characteristics in Agula contributed to a remarkable reduction in runoff coefficient (75%) and increased the BFI between the two periods (Table 7.6). Analysis of hydrological fluxes response between the pre- and post-treatments of the paired catchments behaves consistently (Table 7.6). However, the magnitude of the changes is incomparable between the two catchments. For example, the reduction in surface runoff and runoff coefficients from Agula is more than double compared to Genfel catchment. Similarly, the base flow index in Agula increased by more than 250% whereas the increment in Genfel was only 56%. Such large differences in the magnitude of changes are attributed to the differences in the level of catchment interventions. This indicates catchment management interventions were responsible for most streamflow, evapotranspiration and soil storage change dynamics in the catchments.

Table 7.6: Comparison of rainfall-runoff relationships before and after catchment management interventions (difference is given as post-treatment minus pre-treatment)

Water budget components	Agula catchment			Genfel catchment		
	Pre-treatment	Post-treatment	difference	Pre-treatment	Post-treatment	difference
Annual precipitation (mm/year)	540	642	102	536	634	98
Actual evapotranspiration (mm/year)	222	385	163	214	355	141
Deep Percolation (mm/year)	43	116	73	38	43	5
Change in storage (mm/year)	-9	57	66	-14	33	47
Annual total flow (mm/year)	284	84	-200	298	203	-95
dry season flow (mm/season)	22	37	15	36	27	-9
wet season flow (mm/season)	262	47	-215	262	176	-86
Runoff coefficient (-)	0.53	0.13	-0.40	0.56	0.32	-0.24
Base flow index (BFI)	0.15	0.53	0.38	0.18	0.28	0.10

The results from both paired catchments and pre and post-treatment comparisons demonstrated that the ongoing integrated environmental rehabilitation programs strongly affected the hydrological processes in the region. Increasing vegetation cover over the catchments enhanced actual evapotranspiration and infiltration capacity of the soils (Alemayehu et al., 2009; Chapter 6). At the same time, the introduced physical SWC structures contributed to the reduction of hill-slope runoff and increased concentration time of the flows (Nyssen et al., 2010). The different types of terraces and deep trenches constructed across the slopes that follow the contour of the field enhanced soil infiltration capacity of the catchments. Most of the terraces in the catchment constructed in hillslopes and plateau have significantly reduced overland flow and increased the soil moisture. These structures are the main explanatory candidate for the increased low flow proportion during the dry seasons. At the same time, soil bunds and deep trenches constructed in gentle slopes and agricultural lands have enhanced soil infiltration capacity, reduced peak runoff, and increased groundwater recharge (Huang & Zhang, 2004;

Wang et al., 2013). Generally, the introduced physical SWC structures affected the hydrological regimes which resulted towards a uniform dry-season flow in the catchment.

Our result is in agreement with several previous studies from around the world (Abouabdillah et al., 2014; Schmidt & Zemadim, 2013;Wang et al., 2013; Gates et al., 2011; Lacombe et al., 2008; Mu et al., 2007). These studies evidenced that physical SWC structures made an important contribution in decreasing surface runoff during the peak rainy season and increasing the low flow during the dry months. The overall hydrological processes of a catchment can be modified through the introduction of SWC structures which can change the partitioning of incoming rainfall on the land surface (Gebremeskel et al., 2018; Gates et al., 2011). A number of local studies (Gebremeskel et al., 2018; Guyassa et al., 2016; Haregeweyn et al., 2015; Nyssen et al., 2010; Alemayehu et al., 2009; Descheemaeker et al., 2006b) have also shown that implementation of SWC structures in watersheds resulted in a decrease of surface runoff volume and enhanced availability of water during the dry months. Similarly, some studies (Gebremeskel et al., 2018; Taye et al., 2015; Gebreyohannes et al., 2013) support the finding of this study that groundwater has significantly increased in the previously degraded lands of the region.

7.4.5 Comparison of model parameters between the paired catchments

The fourth approach, i.e. comparison of calibrated model parameters between the paired catchments, was applied to further assess the impact of catchment management on the hydrological behaviour. The value of the most sensitive model parameters was compared to infer if the possible changes in hydrological fluxes are attributed to the surface characteristics of the catchments. The final optimum values of most sensitive parameter sets of the two catchments are presented as supplementary files (Table E-1). The observed differences in the average value of model parameters between the two catchments indicate that the differences in physical catchment characteristics were responsible for the hydrological response variability between the two catchments.

Parameter values related to canopy such as CanopyGapFraction and the ratio of average wet canopy evaporation rate over average precipitation rate (EovR) are proportional for the two catchments, suggesting that there were no significant differences in the vegetation cover improvements between the paired catchments and consequently on the hydrological regimes. This was also demonstrated by the observed small differences in actual evapotranspiration rates between the two catchments (Table 4). In contrast, the values of parameters related to soil and surface characteristics such as saturated hydraulic conductivity (Ksat), infiltration capacity of the soil (InfiltCapSoil), water content at saturation or porosity (thetaS) and soil parameter determining the decrease in Ksat with depth (M) varied between the two catchments. The value of Ksat, InfiltCapSoil, M parameter, and thetaS parameters in Agula are higher than that of Genfel catchment. This suggests that a larger proportion of the input precipitation is contributed to infiltration and groundwater instead of going to direct runoff generation in Agula than in

Genfel catchment. The modified soil and surface parameter values of Agula catchment is plausibly due to the large proportion of physical SWC interventions (Table 7.1).

All changes in parameter values were towards a slow hydrological response in Agula compared to Genfel catchment. Change in model parameters value between the two models signifies that there is a difference in catchment hydrological response behaviour between the two catchments (Gebremicael et al., 2019a; 2013; Tesemma et al., 2010). Seibert & McDonnell (2010) and Gebremicael et al. (2019a) underlined that the comparison of calibrated model parameters value is a powerful tool to distinguish the change in hydrological response of changing environments. However, it should be noted that this method is not straight forward as different parameter values might be equally possible. Within its limitations, the values of calibrated model parameters in the Agula catchment reflects the result obtained from the paired catchment and pre- and post-treatment comparison approaches.

7.5 CONCLUSION AND RECOMMENDATIONS

A series of integrated catchment management interventions have been implemented in the Agula and Genfel catchments since the mid-2000s but in different degrees. This study aimed at understanding the impact of these measures on the overall hydrological processes, particularly the low flow modifications in the catchment. The relationships between the observed flows from before and after the interventions were quantified using different Indicators of Hydrological Alteration parameters and statistical tests. A paired catchment model (i.e. catchment with intensive SWC (Agula) and control with fewer interventions (Genfel) simulation and model-to-model (pre- and post-treatment) comparison were applied to investigate causes of changes in the low flows of the catchments.

The result of this study confirmed that the treated catchment has experienced a significant change in the overall hydrological processes after the implementations of SWC structures. Our study clearly demonstrated that the low flow of Agula catchment increased substantially more than the control catchment. Significant differences in the partitioning of incoming rainfall was observed after the intervention periods. The annual runoff volume in Genfel was greater than Agula by more than 80% after the intervention. This has resulted in a larger difference (60%) of dry period flows between the two catchments. The ratio of base flow to the total discharge in Agula was almost double that of Genfel which explicitly explains that more of the incoming rainfall in Agula contributes to groundwater recharge. This was also ascertained by the observed large differences in the percentage of the dry seasons flow. The annual flow in Agula is distributed more homogeneously among the different seasons compared to Genfel catchment. Furthermore, analysis of low flow between the paired catchments before and after the SWC measures indicates that the remarkable differences in the dry season flows of the two catchments during the post-treatment period were not visible before the rehabilitation programs.

The large differences in the magnitude of changes in the base flow of the two catchments are attributed to the differences in the level of ongoing SWC interventions which have strongly affected the partitioning of incoming precipitation and soil moisture storage. Implementation of physical SWC structures in the catchment contributed to the interception of runoff water and enhanced soil infiltration capacity of the catchments and hence improved water availability in the dry season. However, despite the fact that the low flow in the dry season significantly increased, the total flow of the catchment has declined significantly following implementation of large scale SWC works. The decrease in the total flow of the stream could be partially attributed to the expansion of very small-scale individual irrigation initiatives that may not have been accounted for in the naturalization process because of their small spatial extent, and also to the increase of transpiration in the catchments due to enhanced vegetation cover. This has a negative impact on the availability of water resources to the downstream users. The key finding of this study is that although the SWC works can enhance the availability of water resources at the local level, it may also reduce the downstream flows significantly.

Chapter 8

CONCLUSION AND RECOMMENDATIONS

8.1 CONCLUSION

Improved understanding of hydro-climatic variabilities and associated hydrological processes at different space and time domain is indispensable for enhanced water resources management in the semi-arid Upper Tekeze basin (45,694 km^2). In this basin water resources development efforts by the Ethiopian government have focused on hydropower development, and to a much smaller extent, irrigation. However, land degradation, water scarcity and inefficient utilization of the available water resources continue to be key constraints for development.

The government has given much attention to rehabilitate the degraded lands of the basin by introducing different catchment management interventions. Land cover changed as a result of these interventions, and soil fertility and food production improved. Also surface and groundwater availability changed. However, studies to quantify the impact of catchment management dynamics on the hydrological processes and its consequences for downstream users are limited. Previous studies were focused at either experimental plots or very small watersheds, making it difficult to extrapolate and infer basin-wide implications.

Therefore, the main objective of this thesis is to understand and assess the impact of anthropogenic changes in the landscape on the hydrological processes of the Upper Tekeze basin, and in particular the impact of watershed management interventions. Implementation of watershed management interventions in a catchment may change the rainfall-runoff relationships and understanding their implications at different spatial scales is critical for planning of water and land management programs. The combined use of secondary and primary datasets, remote sensing, laboratory analysis, and hydrological modelling allowed the hydrological processes and their change to be studied.

As ground-based rainfall measurements are sparse and unevenly distributed in the Upper Tekeze basin, the first step of this study was to identify and validate satellite-based rainfall estimates (Chapter 3). Eight products of satellite-based rainfall covering the period 2002 to 2015 were verified against 34 rain gauging stations in the Upper Tekeze basin. These were: CHIRPS, RFEv2, TRMM, CMorph, PERSIANN, CMap, and GPCP. Both point-to-pixel and area-averaged rainfall were used in the comparison. The results indicated that the CHIRPS, RFEv2 and TRMM consistently outperformed other products across all time scales (daily, monthly and seasonally). Relatively high correlation coefficients (>0.5) and low PBIAS (< ±25%) values were obtained compared to other datasets. The performance of all satellite products was lower in the mountainous terrains of the basin, for elevations higher than 2500 m.a.s.l. The CMap, ARC2 and GPCP underestimated rainfall by a large margin across all stations and at basin level.

The performance of all products except TRMM decreased with increasing spatial scales, which may be attributed to systematic errors during interpolation of the observed rainfall over the complex topography of the basin. The poor performance of the satellite estimates in mountainous areas might have also contributed to the overall lower performance when aggregated at the basin scale. The result of this study is a useful reference for future applications of satellite rainfall products, especially in rain gauge sparse and ungauged basins with rugged terrains. It also justifies the utilization of rainfall estimates by CHIRPS over the complex topography of the Upper Tekeze basin, at least at monthly and seasonal timescales.

Next, the relationships between hydrological and climatic trends were analyzed to identify to what extent the hydrological changes in the basin were driven by the climate. This study investigated the trends and change of rainfall and streamflow in the Upper Tekeze basin (Chapter 4).The analyses were carried out for 21 rainfall and 9 streamflow monitoring stations using Mann-Kendall (MK) and Pettitt tests as well as Indicators of Hydrological Alteration (IHA). The results show that rainfall in most meteorological stations experienced neither increasing nor decreasing trends during the dry, short rainy, main rainy seasons and annually, at 95 % confidence level. In contrast, streamflow in most hydrological stations exhibited a decreasing trend in the dry, short, main rainy seasons and annual totals. Findings from both MK and Pettitt tests are consistent for all seasons and stations, but the timing of change points differs for most stations.

The results demonstrate that there is no link between the trends in rainfall and streamflow in the basin. The change in streamflow must therefore be influenced by factors other than rainfall. The significant trends in streamflow could be due to significant changes over time of catchment characteristics, and of land use /land cover (LULC) in particular.

Multi-temporal LULC change and the possible associated drivers were analysed in the source region of the Upper Tekeze basin (Chapter 5). Satellite images, Geographic Information System (GIS) and ground information were combined for the analysis. The general trend observed in the present study is a decrease in vegetation cover and an increase in agricultural and settlement areas in the last four decades. Most LULC categories changed substantially in the last 43 years (1972-2014). Agricultural land continued to grow until 2001 and started to decrease slightly in the 2001-2014 period. A decreasing tendency of agricultural land during the last period is attributed to the rapid expansion of urban areas and homestead forest plantations. Substantial deforestation of natural vegetation cover occurred in the 1972-1989 period and the proportion of degraded and agricultural lands significantly increased during the same period. However, vegetation cover started to recover since the 1990s, when some of the agricultural, bare, and grasslands have turned into vegetation cover.

The most important driving forces for the observed changes were found to be rapid population growth and changing government policies. The recent government policy on environmental rehabilitation and its associated implementation programs have significantly contributed to the recovery of vegetation cover in the region. Strengthening such interventions with high-level participation of local farmers is essential for the sustainability of biophysical resources. Apart from the well-being of the ecosystem and environmental stabilization, much effort is needed to convert the improved vegetation cover into sources of income for the local communities to secure the gains made so far and to prevent further deterioration.

The effect and contribution of LULC changes on the hydrological flows of Geba catchment (5,085 km^2), in the Upper Tekeze basin, was investigated using a combined approach of hydrologic modelling, hydrological alteration analysis and Partial Least Square Regression analysis (Chapter 6). The human-induced landscape transformations described and analysed in Chapter 5 were used for the development of a spatially distributed hydrological model based on the Wflow-PCRaster/Python modelling framework. This model was used to simulate the hydrological processes and investigate the impact of human-induced environmental changes on the streamflow of the basin. The results show that the expansion of agricultural and grazing land at the expense of natural vegetation cover during the period 1972 to 1989 increased surface runoff and contributed to a decrease of the dry season flow. Interestingly, in the period of 1989 to 2001 dry season flows increased again and surface runoff decreased due to improvements in natural vegetation cover from the mid-1990s. The wet season flows generated from surface runoff continued to decline in the most recent period (2001-2014), but this was accompanied by an unexpected decline in dry season flow as well, which was attributed to an increase in water withdrawals for irrigation.

These findings are in agreement with the analysis of hydrological alteration parameters where the change in magnitude of median monthly flow, annual extremes, frequency and duration of flow pulses and rate and frequency of water conditions showed a change between the periods. The hydrological response to LULC change was more pronounced at sub-catchment level, which is mainly linked to the observed uneven spatial distribution of degraded lands and subsequent rehabilitated lands. This study has unequivocally demonstrated that the ongoing watershed management interventions can increase dry season flows, while decreasing wet season flows at catchment scale (~5,000 km^2). This finding is good news and salient for stakeholders as dry season flows occur when water is most needed.

Finally, to complete the general picture on the impact of integrated catchment management on the hydrological processes, the low-flow responses to physical soil and water conservation interventions was investigated at sub-catchment scale (~500 km^2) using a paired catchment modelling approach (Chapter 7). A paired catchment comparison approach (one with intensive SWC interventions and the other as control with significantly fewer SWC interventions) and a model-to-model (pre- and post-treatment) comparison approach using the model developed in Chapter 6 were applied to understand the responses to physical SWC interventions in more detail. Furthermore, hydrological alteration indicators were used to verify if the hydrological regime of a catchment changes due to SWC intervention programs. The results clearly demonstrated that a substantial increase of the low-flow was observed in the treated catchment compared to the control catchment. Despite the very close similarities in the other physical characteristics of the two adjacent catchments, significant differences in the partitioning of incoming rainfall were observed after the intervention. The annual runoff volume in the treated (with SWC interventions) catchment was lower than that of the untreated (with significantly fewer SWC interventions) catchment by more than 80%. The ratio of base flow to total discharge in the treated catchment was almost double that of the untreated catchment, demonstrated that more of the incoming rainfall in the treated catchment contributes to groundwater recharge. The annual flow in the treated catchment was distributed more homogeneously among the different seasons compared to the untreated catchment.

The large differences in the magnitude of changes in the base flow of the two catchments are attributed to the differences in the level of ongoing SWC interventions which have strongly affected the partitioning of incoming precipitation and soil moisture storage. Implementation of physical SWC structures contributed to the interception of runoff water and enhanced soil infiltration capacity and hence improved water availability in the dry season. Despite the fact that the low-flow in the dry season significantly increased, the total flow of the catchment showed a declining trend. The decrease of total streamflow is attributed to the expansion of small scale irrigation schemes and increase in evapotranspiration from increased vegetation cover. This negatively impacts the availability of water resources for downstream users.

The main findings of this study is that human-induced environmental changes are the most important causes of change in hydrological processes and availability of water resources in the Upper Tekeze basin. Climate change (in the form of precipitation) has, to date, not significantly impacted the availability of water resources in the basin. This thesis has shown that the ongoing watershed management interventions in the Upper Tekeze basin have impacted water availability differently at different spatial and temporal scales. Table 8.1 summarizes the streamflow response before and after watershed management interventions at different spatial scales, namely at 500, 5,000 and 50,000 km^2. Significant changes in the magnitude of streamflow components (annual totals, wet and dry season flows) were found at all spatial scales. While rainfall slightly increased, the wet season flow decreased and dry season flow increased at all three spatial scales after the intensive watershed management intervention programs in the headwater catchments of the basin. Note that the observed large increase in dry season flow and large decrease in wet season flow at the basin outlet (Embamadre) is only partially due to the effect of upstream watershed management interventions, but may largely be attributed to the Tekeze hydropower dam 80 km upstream this station, which was inaugurated in 2009. The large decrease in total annual flow at the small and medium scale may largely be attributed to water withdrawals for smallholder irrigation farming.

The rate of change in streamflow over larger catchments appear to be smaller compared to the smaller catchments. This typical scale effect is mainly associated with the uneven spatial distribution of watershed management interventions in sub-catchments. Enhanced dry season flows and reduced surface runoff from sub-catchments with watershed management interventions are counter-balanced by a decrease in dry season flows and increased in wet season flows from sub-catchments without such interventions. In conclusion, the collective evidence from this study confirms that the impact of integrated catchment management interventions on the availability of water is most pronounced at local level.

Table 8.1: Effect of watershed management interventions (before and after intervention) on the rainfall-runoff relationships at different spatial scales.

Flow components	Embamadre (A ≈ 50,000 km²)			Geba (A ≈ 5,000 km²)			Agula (A ≈ 500 km²)		
	Before (2003-06)	After (2014-16)	Change	Before (2003-06)	After (2013-15)	Change	Before (2003-06)	After (2014-16)	Change
Annual precipitation (mm/year)	864	903	39	586	600	14	540	642	102
Annual total flow (mm/year)	205	234	29	213	141	-72	284	84	-200
dry season flow (mm/season)	19	87	68	47	63	16	22	37	15
wet season flow (mm/season)	180	138	-42	150	94	-56	262	47	-215
Runoff coefficient (-)	0.23	0.25	0.02	0.36	0.24	-0.12	0.53	0.13	-0.4
Base flow index (BFI)	0.21	0.43	0.22	0.24	0.47	0.23	0.15	0.53	0.38

In conclusion, the collective evidence from this studies confirms that the impact of integrated catchment management interventions on the availability of water is more pronounced at local level. This study has also proved that although the SWC works can enhance the availability of water resources during the dry season at the local level, it may reduce the total annual flow downstream. This finding entails that implementation of catchment management strategies should carefully consider their positive and negative impacts both for upstream and downstream stakeholders. The approach employed in this thesis has unequivocally demonstrated the impact of land management interventions on the hydrology of the basin, at different temporal and spatial scales. In summary, this study provides the following scientific contributions

✓ Provides an improved understanding on the performance of different satellite rainfall products and their applications over a complex topography and variable climate

✓ Improved understanding on the drivers of streamflow variabilities and changes in semi-arid catchments

✓ Improved understanding on the relationships of hydrological processes and environmental changes in complex catchment characteristics

✓ There was no distinct understanding in the literature on how watershed management interventions affects hydrological processes in semi-arid catchments. This study has provided a clear understanding on how large scale implementations of these interventions can influence the overall hydrological response, in particular the low flows of a catchments.

✓ Development of a distributed hydrological model in PCRaster/Python modelling framework for catchments with complex characteristics can improve on understanding the impact of human interventions on the hydrological processes and variabilities.

✓ An integration of rainfall-runoff modelling, indicators of hydrological alteration parameters and PLSR analysis is found to be robust to assess the impact of environmental change on the hydrology of complex catchments.

It is also essential to point out some limitations of this study. Poor quality and limited coverage of hydro-climatic data across the basin may have introduced some uncertainties in the results. The length of the record period used for the trend analyses of hydro-climatic data varies across all monitoring stations and these different record lengths of data may have introduced some discrepancies. Uncertainties and inconsistencies from data collection devices (sensors), errors in LULC class definitions and uneven distribution of reference data may also have affected the hydrological modelling study in Chapters 6 and 7.

8.2 RECOMMENDATIONS

The influence of human-induced environmental changes on the hydrological processes of the Upper Tekeze basin were quantified at different spatial and temporal scales. This thesis provides an in-depth understanding of the rainfall-runoff relationship and streamflow variability in response to the different catchment management interventions in the basin. The following recommendations are outlined for further research to improve land and water resources management.

1. Despite of the availability of remote sensing products, the few meteorological stations that are operational should be maintained and if needed improved, ensuring that they continue to collect high quality data, with which new products can be evaluated.

2. Trend analysis and change point detection of rainfall and streamflow can be used to justify studying catchment processes and the impact of catchment management dynamics on the hydrological processes in other basins. However, this requires long term and reliable data sets.

3. Improved understanding of land use/cover change dynamics allows more reliable information for improved understanding of hydrological processes in catchments. The information on the land use/cover dynamics provided in this study can be used for further hydrological analysis and designing of improved land management interventions in the region

4. This study showed that the implementation of catchment management interventions improved water availability in the dry season at the local level. However, the total discharge

out of the catchment declined significantly following implementation of large scale SWC works. This negatively impacts the availability of water resources for downstream users, including transboundary stakeholders. Further research is needed to identify effective catchment management strategies that can balance the availability of, and access to, green and blue water resources at both local and larger scales.

5. SWC interventions that increase dry season flows can be considered a nature-based solution reducing the need for physical storage structures. A cost-benefit analysis on the implementation of large scale catchment management interventions versus building water storage dams is needed to better understand the economic trade-offs between these interventions.

6. The results from this study suggests that catchment managers and decision-makers should pay attention to the equitable and sustainable utilization, at the local level, of the enhanced water resources because of SWC interventions.

7. The hydrological modelling approach applied in this study was found to be capable to quantify hydrological responses to human-induced environmental change in a complex catchment. The development of a fully distributed hydrological model in the wflow_PCRaster/Python modelling framework has potential to simulate all salient hydrological components by maximizing available spatial data with little calibration to minimize the risks associated with over-parameterization.

REFERENCES

Abbott, M. B., Bathurst, J. C., Cunge, J. A., O'Connell, P. E., & Rasmussen, J.: An introduction to the European Hydrological System-Systeme Hydrologique Europeen, "SHE": Structure of a physically-based, distributed modelling system, J. Hydrol., 87(1–2), 61-77, 1986a.

Abdi, H.: Partial least squares regression and projection on latent structure regression (PLS R). Interdiscip. Rev. Comput. Stat. 2, 97-106, 2010.

Abdul Aziz, O. I., and Burn, D. H.: Trends and variability in the hydrological regime of the Mackenzie River Basin, J. Hydrol., 319, 282-294, 2006.

Abeysingha, N. S., Singh, M., Sehgal, V. K., Khanna, M., and Pathak, H.: Analyses of trends in streamflow and its linkages with rainfall and anthropogenic factors in Gomti River basin of North India, Theor. Appl. Climatol., 123, 785-799, 2015.

Abouabdillah, A., White, M., Arnold, J., De Girolamo, A., Oueslati, O., Maataoui, A., & Lo Porto, A.: Evaluation of soil and water conservation measures in a semi-arid river basin in Tunisia using SWAT, Soil Use and Manag., 30(4), 539-549, 2014.

Abraha, A., Ashenafi, A.: Modeling Hydrological Responses to Changes in Land Cover and Climate in Geba River Basin, Northern Ethiopia, PhD, Freie Universität Berlin, Berlin, Germany, 2014.

Al-Ahmadi, F., & Hames, A. S.: Comparison of Four Classification Methods to Extract Land Use and Land Cover from Raw Satellite Images for Some Remote Arid Areas, Kingdom of Saudi Arabia, JKAU; Earth Sci., 20(1), 167-191, 2009.

Alemayehu, F., Taha, N., Nyssen, J., Girma, A., Zenebe, A., Behailu, M., Poesen, J.: The impacts of watershed management on land use and land cover dynamics in Eastern Tigray (Ethiopia), Resour. Conserv. Recy. 53(4), 192-198, 2009.

Ampadu, B., Chappell, N. A., & Kasei, R. A.: Rainfall-riverflow modelling approaches: making a choice of data-based mechanistic modelling approach for data limited catchments: a review, CJPAS, 7(3), 2571-2580, 2013.

Amsalu, A., Stroosnijder, L., & Graaff, J.: Long-term dynamics in land resource use and the driving forces in the Beressa watershed, highlands of Ethiopia, J. Environ. Manag., 83(4), 448-459, 2007.

Anderson, B. J. R., Hardy, E. E., Roach, J. T., & Witmer, R. E.: A land use and landcover classification system for use with remote sensor data, Devel., 2005,, 28-28, 1979.

Ariti, A. T., van Vliet, J., & Verburg, P. H.: Land-use and land-cover changes in the Central Rift Valley of Ethiopia: Assessment of perception and adaptation of stakeholders, Appl. Geogr. 65, 28-37, 2015.

Ariti, A. T., van Vliet, J., Verburg, P. H.: Land-use and land-cover changes in the Central Rift Valley of Ethiopia: Assessment of perception and adaptation of stakeholders, Appl. Geogr. 65, 28-37, 2015.

Arnold, J., Srinivasan R, Muttiah, R., Williams J.: Large area hydrologic modeling and assessment: Part I. Model development, JAWRA, 34(1): 73–89, 1998.

Asadullah, A., N. McIntyre, and Kigobe, M.: Evaluation of Five Satellite Products for Estimation of Rainfall over Uganda, Hydrol. Sci. J., 53: 1137-1150, 2008.

Asfaha, T. G., Frankl, A., Haile, M., & Nyssen, J.: Catchment rehabilitation and hydro-geomorphic characteristics of mountain streams in the western rift valley escarpment of northern Ethiopia, Land Degrad. Dev. 27, 26–34, 2014.

Ashouri, H. S. Hsu, D. Sorooshian, K. Braithwaite, B. Knapp, B. Cecil, and C. Prat.: PERSIANN-CDR: Daily Precipitation Climate Data Record from Multisatellite Observations for Hydrological and Climate Studies, BAMS., 96: 69-83, 2015.

Awulachew, S. B., Yilma, A. D., Loulseged, M., Loiskandl, W., Ayana, M., and Alamirew, T.: Water resources and irrigation development in Ethiopia, IWMI, Colombo, Sri Lanka, 2007.

Awulachew, S. B.: Irrigation potential in Ethiopia: Constraints and opportunities for enhancing the system, IWMI Working paper, Colombo, Sri Lanka, 2010.

Ayenew, T.: Water management problems in the Ethiopian rift: Challenges for development, *J. Afr.Earth Sci., 48*(2–3), 222-236, 2007.

Balthazar, V., Vanacker, V., Girma, A., Poesen, J., & Golla, S.: Human impact on sediment fluxes within the Blue Nile and Atbara River basins. *Geomorphology, 180*, 231-241, 2013.

Bastiaanssen, W. G. M., Menenti, M., Feddes, R. A., & Holtslag, A. A. M.: A remote sensing Surface Energy Balance Algorithm for Land (SEBAL) 1, Formulation, *J. Hydrol,, 212-213*, 198-212, 1998a.

Bastiaanssen, W., Noordman, E., Pelgrum, H., Davids, G., Thoreson, B., & Allen, R.: SEBAL model with remotely sensed data to improve water-resources management under actual field conditions, *J. Irrig. Drain. Eng., 131*(1), 85-93, 2005.

Batelaan, O., and De Smedt, F.: WetSpass: A flexible, GIS based, distributed recharge methodology for regional groundwater modelling. In Impact of Human Activity on Groundwater Dynamics, ed. H.Gehrels, J.Peters, E.Hoehn, K.Jensen, C.Leibundgut, 2001.

Bayissa, Y., T. Tadesse, G. Demisse, and A. Shiferaw.: Evaluation of Satellite-Based Rainfall Estimates and Application to Monitor Meteorological Drought for the Upper Blue Nile Basin, Ethiopia, *Remote Sens.* 9: 669, 2017.

Behrangi, A., K. Behnaz, C. Tsou, A. Amir, Kuolin, S. Soroosh, and N. Bacchetta.: Hydrologic Evaluation of Satellite Precipitation Products over a Mid-Size Basin, *J. Hydrol.,* 397: 225-237, 2015.

Belay, K. T., Van Rompaey, A., Poesen, J., Van Bruyssel, S., Deckers, J., and Amare, K.: Spatial Analyses of Land Cover Changes in Eastern Tigray (Ethiopia) from 1965 to 2007: Are There Signs of a Forest Transition?, *Land Degrad. Devel.* 26, 680-689, 2015.

Belete, k.: Sedimentation and Sediment Handling at Dams in Tekeze River Basin, Ethiopia, (PhD thesis), Norwegian University of Science and Technology, Trondheim, Norway, 2007.

Bergkamp, G.: A hierarchical view of the interactions of runoff and infiltration with vegetation and microtopography in semiarid shrublands, *CATENA, 33*(3–4), 201-220, 1998.

Best A, Zhang L, McMahon T, Western A, Vertessy R.: A critical review of paired catchment studies with reference to seasonal flows and climate variability, cooperative research centre for catchment hydrology report02/x. Monash University, Victoria, Australia, 2002.

Beven, K.: Rainfall-runoff modelling: the primer. Chichester, England: John Wiley & Sons, 2011.

Beven, K.: Towards an alternative blueprint for a physically-based digitally simulated hydrologic response modelling system, *Hydrol. Process., 16*(2), 189-206, 2002.

Bewket, W., & Sterk, G.: Dynamics in land cover and its effect on stream flow in the Chemoga watershed, Blue Nile basin, Ethiopia. *Hydrol. Process. 19*, 445-458. 2005.

Bharti, V., and Singh, C.: Evaluation of error in TRMM 3B42V7 precipitation estimates over the Himalayan region, *J. Geophys. Res-Atmos.*, 120, 12458-12473, 2015.

Biazin, B., & Sterk, G.: Drought vulnerability drives land-use and land cover changes in the Rift Valley dry lands of Ethiopia, *Agric. Ecosyst. Environ. 164*, 100-113, 2012.

Bishaw, B.: Deforestation and land degradation in the Ethiopian highlands: a strategy for physical recovery, *NEAS*, 7-25, 2001.

Bishop, P. L., Hively, W. D., Stedinger, J. R., Rafferty, M. R., Lojpersberger, J. L., & Bloomfield, J. A.: Multivariate analysis of paired watershed data to evaluate agricultural best management practice effects on stream water phosphorus, *J.Environ.Qual.*, 34(3), 1087-1101, 2005.

Bizuneh, A.: Modeling the Effect of Climate and Land Use Change on theWater Resources in Northern Ethiopia: the Case of SuluhRiver Basin, (PhD), Freie Universität, Berlin, Germany., 2013.

Bradford, M. J., & Heinonen, J. S.: Low flows, instream flow needs and fish ecology in small streams, *Can Water Resou. J.*, 33(2), 165-180, 2008.

Braimoh, A. K.: Random and systematic land-cover transitions in northern Ghana, *Agric. Ecosyst. Environ.* 113, 254-263, 2006.

Brown, A. E., Zhang, L., McMahon, T. A., Western, A. W., & Vertessy, R. A.: A review of paired catchment studies for determining changes in water yield resulting from alterations in vegetation, *J. Hydrol.*, 310(1), 28-61, 2005.

Burn, D., Cunderlik, J. M., and Pietroniro,A.: Hydrological trends and variability in the Liard River basin, *Hydrol. Sci. J.*, 49, 53-67, 2004.

Camacho, V., Saraiva Okello, A., Wenninger, J., & Uhlenbrook, S.: Understanding runoff processes in a semi-arid environment through isotope and hydrochemical hydrograph separations, *Hydrol. Earth Sys. Sci.*, 12(1), 975-1015, 2015.

Cammeraat, E. L. H. (2004). Scale dependent thresholds in hydrological and erosion response of a semi-arid catchment in southeast Spain, *Agri. Ecosys. Environ.*, 104(2), 317-332.

Carver and Nash, R.H.: Doing data analysis with SPSS version 16, Cengage Learning, Belmont, 2009.

Castiglioni, S., Castellarin, A., Montanari, A., Skøien, J., Laaha, G., & Blöschl, G.: Smooth regional estimation of low-flow indices: physiographical space based interpolation and top-kriging.,*Hydrol. Earth Sys. Sci.*, 15(3), 715-727, 2011.

Castillo, V. M., Gómez-Plaza, A., & Martínez-Mena, M.: The role of antecedent soil water content in the runoff response of semiarid catchments: a simulation approach, *J. Hydrol.*, 284(1–4), 114-130, 2004.

Castillo, V. M., Gómez-Plaza, A., and Martínez-Mena, M.: The role of antecedent soil water content in the runoff response of semiarid catchments: a simulation approach, *J. Hydrol.*, 284, 114-130, 2003.

Chai. T and Draxler. R. R.: Root mean square error (RMSE) or mean absolute error (MAE)? Arguments against avoiding RMSE in the literatur, *Geosci. Model Dev.*, 7, 1247–1250, 2014.

Chang, J., Zhang, H., Wang, Y., & Zhu, Y.: Assessing the impact of climate variability and human activities on streamflow variation, *Hydrol.Earth Sys. Sci.*, 20(4), 1547-1560, 2016.

Cheema, M., & Bastiaanssen, W. G.: Land use and land cover classification in the irrigated Indus Basin using growth phenology information from satellite data to support water management analysis, *Agric.Water. Manage.* 97(10), 1541-1552., 2010.

Chen, Q., Mei, K., Dahlgren, R. A., Wang, T., Gong, J., & Zhang, M.: Impacts of land use and population density on seasonal surface water quality using a modified geographically weighted regression, *Sci. Total Environ.*, 572, 450-466, 2016.

Congalton, G.: Accuracy assessment and validation of remotely sensed and other spatial informa, *Int. J. Wildland Fire*, 10, 321-328, 2001.

Congcong, L., Wang, J., Wang. L, Luanyun. H, & Peng.G.: Comparison of Classification Algorithms and Training Sample Sizes in Urban Land Classification with Landsat Thematic Mapper Imagery, *Remote Sens. 6*, 964-983, 2014.

Conway, D., and Hulme, M.: Recent fluctuations in precipitation and runoff over the Nile sub-basins and their impact on main Nile discharge, *Climatic Change*, 25, 127-151, 1993.

CSA.: Summary and statistical report of the 2007 population and housing census results. Population size by age and sex. Central Statistical Agency of Ethiopia, Addis Abeba, Ethiopia, 2008.

CSA.: The 1994 population and housing census of Ethiopia. Results for Tigray region. Central Statistical Agency of Ethiopia,Addis Ababa, Ethiopia, 1994.

Daniel, B., Camp,V., LeBoeuf, J., Penrod, R., Dobbins, J. P., & Abkowitz, M. D.: Watershed modeling and its applications: A state-of-the-art review, *Open Hydrol. J.*, 5(2), 25-50, 2011.

Das, T., Bárdossy, A., Zehe, E., & He, Y.: Comparison of conceptual model performance using different representations of spatial variability, *J. Hydrol.*, *356*(1), 106-118, 2008.

De Wit A.: Runoff controlling factors in various sized catchments in a semi-arid Mediterranean environment in Spain. (PhD), Utrecht University, Utrecht, The Netherlands, 2001.

Degefu, M, A., P. R. David, and W. Bewket.: Teleconnections between Ethiopian Rainfall Variability and Global SSTs: Observations and Methods for Model Evaluation, *Meteorol. Atmos. Phys.*, 129: 173–182, 2017.

Dembélé, M., and S.J. Zwart.: Evaluation and Comparison of Satellite-Based Rainfall Products in Burkina Faso, West Africa. *Int. J.Remote Sens.*, 37: 3995-4014, 2016.

Derin, Y., and K. K. Yilmaz.: Evaluation of Multiple Satellite-Based Precipitation Products over Complex Topography, *J. Hydrometeorol.*, 15: 1498-1516, 2014.

Descheemaeker, K., Muys, B., Nyssen, J., Poesen, J., Raes, D., Haile, M., & Deckers, J.: Litter production and organic matter accumulation in exclosures of the Tigray highlands, Ethiopia. *For. Ecol. Manag. 233*(1), 21-35, 2006a.

Descheemaeker, K., Nyssen, J., Poesen, J., Raes, D., Haile, M., Muys, B., Deckers, S.: Runoff on slopes with restoring vegetation: A case study from the Tigray highlands, Ethiopia. *J. Hydrol.* 331, 219-241, 2006b.

Descheemaeker, K., Nyssen, J., Rossi, J., Poesen, J., Haile, M., Raes, D., Muys, B., Moeyersons, J., and Deckers, S.: Sediment deposition and pedogenesis in exclosures in the Tigray highlands, Ethiopia, *Geoderma*, 132, 291-314, 2006c.

Descheemaeker, K., Poesen, J., Borselli, L., Nyssen, J., Raes, D., Haile, M., Muys, B., and Deckers, J.: Runoff curve numbers for steep hillslopes with natural vegetation in semi-arid tropical highlands, northern Ethiopia, *Hydrol. Process.*, 22, 4097-4105, 2008.

Dinku, T., Ceccato, P., Grover-Kopec, E., Lemma, M., Connor, S., and Ropelewski, C.: Validation of satellite rainfall over East Africa's complex topography, *Int. J. Remote Sens.*, 28, 1503-1526, 2007.

Dixon, H., Lawler, D. M., Shamseldin, A. Y., Webster, P., Demuth, S., Gustard, A., Planos, E., Seatena, F., and Servat, E.: The effect of record length on the analyses of river flow trends in Wales and central England, Climate variability and change: hydrol. *Impacts*, 490-495, 2006.

Feidas, H.: Validation of Satellite Rainfall Products over Greece, *Theor. Appl. Climatol.*, 99: 193-216, 2010.

Fenta, A.A., Yasuda, H., Shimizu, K., Haregeweyn, N., 2017. Response of streamflow to climate variability and changes in human activities in the semiarid highlands of northern Ethiopia. *Reg. Environ. Change.*, 17, 1229-1240.

Foody, G.: Areal photographs, previous local and regional studies and an in-depth interviews, *Remote. Sens. Environ.*, 80, 185-201, 2002.

Frankl, A., Poesen, J., Deckers, J., Haile, M., & Nyssen, J.: Gully head retreat rates in the semi-arid highlands of Northern Ethiopia. *Geomorphology*, 173–174(0), 185-195, 2012.

Funk, C. C., P. J. Peterson, M. F. Landsfeld, D. H. Pedreros, P. Verdin, J. D. J. Rowland, and A. P. Verdin.: A Quasi-Global Precipitation Time Series for Drought Monitoring. US Geological Survey Data Series, 832, 2014.

Gao, P., Mu, X.-M., Wang, F., and Li, R.: Changes in streamflow and sediment discharge and the response to human activities in the middle reaches of the Yellow River, *Hydrol. Earth Syst. Sci.*, 15, 1-10, 2011.

Garg, K. K., Karlberg, L., Barron, J., Wani, S. P., & Rockstrom, J.: Assessing impacts of agricultural water interventions in the Kothapally watershed, Southern India, *Hydrolo. Proces., 26*(3), 387-404, 2012.

Gash, J. H. C., Lloyd, C. R., & Lachaud, G.:. Estimating sparse forest rainfall interception with an analytical model, *J. Hydrol., 170*(1), 79-86, 1995.

Gashaw, T., Tulu, T., Argaw, M., & Worqlul, A. W.: Modeling the hydrological impacts of land use/land cover changes in the Andassa watershed, Blue Nile Basin, Ethiopia, *Sci. Total Environ., 619*, 1394-1408, 2018.

Gates, J. B., Scanlon, B. R., Mu, X., & Zhang, L.: Impacts of soil conservation on groundwater recharge in the semi-arid Loess Plateau, China, *Hydrogeol. J., 19*(4), 865-875, 2011.

Gebrehiwot, T., van der Veen, A., and Maathuis, B.: Spatial and temporal assessment of drought in the Northern highlands of Ethiopia, *Int. J. Appl. Earth Obs.*, 13, 309-321, 2011.

Gebremedhin, A. Mefin., H., Abrah, A., Abraha, G., Misgina, S., Gebremicael, T.G.: Impact of Climate Change on net Irrigation Water Requirement of major crops in the semi-arid regions of Northern Ethiopia, *Journal of the Drylands*, 8(1), 729-740, 2018.

Gebremeskel, G., Gebremicael, T. G., Girmay, A.: Economic and environmental rehabilitation through soil and water conservation, the case of Tigray in northern Ethiopia, *J. Arid Environ.* 151, 113-124, 2018.

Gebremeskel, G., Kebede, A.: Estimating the effect of climate change on water resources: Integrated use of climate and hydrological models in the Werii watershed of the Tekeze river basin, Northern Ethiopia. *Agriculture and Natural Resources* 52, 195-207, 2018.

Gebremicael, T. G., Mohamed, Y. A., & Van der Zaag, P.: Attributing the hydrological impact of different land use types and their long-term dynamics through combining parsimonious hydrological modelling, alteration analysis and PLSR analysism, *Sci. Total Environ., 660*, 1155-1167, 2019a.

Gebremicael, T. G., Mohamed, Y. A., Betrie, G. D., van der Zaag, P., and Teferi, E.: Trend analyses of runoff and sediment fluxes in the Upper Blue Nile basin: A combined analyses of statistical tests, physically based models and landuse maps, *J.Hydro.*, 482, 57-68, 2013.

Gebremicael, T. G., Mohamed, Y. A., van der Zaag, P., & Hagos, E. Y.: Temporal and spatial changes of rainfall and streamflow in the Upper Tekeze–Atbara River Basin, Ethiopia. *Hydrol. Earth. Syst. Sci.* 21, 2127-2142, 2017.

Gebremicael, T. G., Mohamed, Y. A., van der Zaag, P., Hagos, E.: Quantifying longitudinal land use change from land degradation to rehabilitation in the headwaters of the Tekeze-Atbara basin, Ethiopia. *Sci.Total Environ.* 622-623, 1581-1589, 2018.

Gebremicael, T. G., Mohamed, Y. A., Van der Zaag, P., Gebremedhin, A., Gebremeskel, G., Yazew, E., & Kifle, M.: Evaluation of multiple satellite rainfall products over the rugged topography of the Tekeze-Atbara basin in Ethiopia, *Int. J. Remote Sens.*, 1-20, 2019b.

Gebremichael, M., M. M. Bitew, F. A. Hirpa, and G. N. Tesfay.: Accuracy of Satellite Rainfall Estimates in the Blue Nile Basin: Lowland Plain Versus Highland Mountain, *Water Res. Res.*, 50: 8775-8790, 2014.

Gebresamuel, G., Bal, R. S., & Øystein, D.: Land-use changes and their impacts on soil degradation and surface runoff of two catchments of Northern Ethiopia. *Acta Agr Scand B-S P, 60*(3), 211-226, 2010.

Gebreyohannes, T., De Smedt, F., Walraevens, K., Gebresilassie, S., Hussien, A., Amare, K., Deckers, J., and Gebrehiwot, K.: Application of a spatially distributed water balance model for assessing surface water and groundwater resources in the Geba basin, Tigray, Ethiopia, *J. Hydro.*, 499, 110-123, 2013.

Girmay, G., Singh, R., Nyssen, J., & Borrosen, T.: Runoff and sediment-associated nutrient losses under different land uses in Tigray, Northern Ethiopia, *J.Hydrol.* 376,70-80, 2009.

Githui, F. W.: Assessing the impacts of environmental change on the hydrology of the Nzoia catchment, in the Lake Victoria Basin. (PhD), Vrije Universiteit Brussel, Brussels, Belgium, 2008.

Giuntoli, I., Renard, B., Vidal, J. P., & Bard, A.: Low flows in France and their relationship to large-scale climate indices, *J. Hydrol.*, 482, 105-118, 2013.

Goitom, H.: Modeling of Hydrological Processes in the Geba River Basin, Northern Ethiopia. (PhD), Vrije Universiteit Brussel, Belgium, 2012.

Goovaerts, P. Geostatistical approaches for incorporating elevation into the spatial interpolation of rainfall, *J. Hydrol.*, 228, 113-129, 2000.

Grieser, J., Gommes, R., and Bernardi, M.: New LocClim–the local climate estimator of FAO, *Geophysical Research Abstracts*, 8, 0305, 2006.

Gumindoga, W., Rwasoka, D., & Murwira, A.: Simulation of streamflow using TOPMODEL in the Upper Save River catchment of Zimbabwe, *Phys.Chem. Earth.*, 36, 806-813, 2011.

Guo, R., and Y. Liu.: Evaluation of Satellite Precipitation Products with Rain Gauge Data at Different Scales: Implications for Hydrological Applications, *Water* 8: 28, 2006.

Guyassa, E., Frankl, A., Zenebe, A., Poesen, J., & Nyssen, J.: Effects of check dams on runoff characteristics along gully reaches, the case of Northern Ethiopia, *J. Hydrol.*, 545, 299-309, 2016.

Guyassa, E., Frankl, A., Zenebe, A., Poesen, J., & Nyssen, J.: Gully and soil and water conservation structure densities in semi-arid northern Ethiopia over the last 80 years, *Earth Surf Process Landf,* 2018.

Guzha, A. C., Rufino, M. C., Okoth, S., Jacobs, S., & Nóbrega, R. L. B.: Impacts of land use and land cover change on surface runoff, discharge and low flows: Evidence from East Africa, *EJRH*, 15, 49-67, 2018.

Gyamfi, C., Ndambuki, J. M.,Salim, R. W.: Hydrological Responses to Land Use/Cover Changes in the Olifants Basin, South Africa. *Water*, 8, 588, 2016.

Haile, A., E. Habib, and T. Rientjes. 2013. Evaluation of the Climate Prediction Center (CPC) Morphing Technique (CMORPH) Rainfall Product on Hourly Timescales over the Source of the Blue Nile Rive, *Hydrol. Process.*, 1829-1839, 2013.

Haile, A.T., T. Rientjes, A. Gieske, and M. Gebremichael.: Multispectral Remote Sensing for Rainfall Detection and Estimation at the Source of the Blue Nile River, *Int J Appl Earth Obs Geoinf*, 12: S76-S82, 2010.

Hannaford, J.: Climate-driven changes in UK river flows: A review of the evidence, *Prog. Phys. Geogr.*, 39, 29-48, 2015.

Haregeweyn, N., Poesen, J., Nyssen, J., De Wit, J., Haile, M., Govers, G., & Deckers, S.: Reservoirs in Tigray (Northern Ethiopia): characteristics and sediment deposition problems, *Land. Degrad. Dev.* 17(2), 211-230, 2006.

Haregeweyn, N., Tesfaye, S., Tsunekawa, A., Tsubo, M., Meshesha, D., Adgo, E., & Elias, A.: Dynamics of land use and land cover and its effects on hydrologic responses: Gilgel Tekeze catchment in the highlands of Northern Ethiopia, *Environ. Monit. Assess. 187*, 1-14, 2014.

Haregeweyn, N., Tesfaye, S., Tsunekawa, A., Tsubo, M., Meshesha, D., Adgo, E., Elias, A.: Dynamics of land use and land cover and its effects on hydrologic responses: case study of the Gilgel Tekeze catchment in the highlands of Northern Ethiopia. *Environ., Monit. Assess.* 187, 1-14, 2014.

Haregeweyn, N., Tsunekawa, A., Nyssen, J., Poesen, J., Tsubo, M., Tsegaye Meshesha, D., Schütt, B., Adgo, E., and Tegegne, F.: Soil erosion and conservation in Ethiopia: A review, *Prog. Physic. Geog.* 39, 750-774. 2015.

Haregeweyn, N., Tsunekawa, A., Nyssen, J., Poesen, J., Tsubo, M., Tsegaye Meshesha, D.,Tegegne, F.: Soil erosion and conservation in Ethiopia: A review, *Prog. Phys. Geog.,* 39(6), 750-774, 2015.

Haregeweyn, N., Tsunekawa, A., Poesen, J., Tsubo, M., Meshesha, D.T., Fenta, A.A., Nyssen, J., Adgo, E.: Comprehensive assessment of soil erosion risk for better land use planning in river basins: Case study of the Upper Blue Nile River. *Sci.Total Environ.*, 574, 95-10, 2017.

Hargreaves, G. H., Riley, J. P.: Agricultural benefits for Senegal River basin, *J. Irrig. Drain. Eng.* 111(2), 113-124, 1985.

Hassaballah, K., Mohamed, Y., Uhlenbrook, S., Biro, K.: Analysis of streamflow response to land use and land cover changes using satellite data and hydrological modelling: case study of Dinder and Rahad tributaries of the Blue Nile (Ethiopia–Sudan), *Hydrol. Earth Syst. Sci.* 21, 5217, 2017.

Hassan, Z., Shabbir, R., Ahmad, S. S., Malik, A. H., Aziz, N., Butt, A., & Erum, S.: Dynamics of land use and land cover change (LULCC) using geospatial techniques: a case study of Islamabad Pakistan., *SpringerPlus*, 5, 812, 2016.

He, Z., L. Yang, T. Fuqiang, N. Guangheng, H. Aizhong, and L. Hui.: Intercomparisons of Rainfall Estimates from TRMM and GPM Multisatellite Products over the Upper Mekong River Basin, *J. Hydrometeorol.,* 18: 413-430, 2017.

Hengl, T., de Jesus, J. M., Heuvelink, G. B., Gonzalez, M. R., Kilibarda, M., Blagotić, A.,Bauer-Marschallinger, B.: SoilGrids250m: Global gridded soil information based on machine learning, *PLoS One*, 12, e0169748, 2017.

Hengsdijk, H., Meijerink, G. W., & Mosugu, M. E.: Modeling the effect of three soil and water conservation practices in Tigray, Ethiopia, *Agri. Ecosy. Environ.,* 105(1-2), 29-40, 2005.

Herman, A. P. Kumar, P. Arkin, and V. Kousky.: Objectively Determined 10-Day African Rainfall Estimates Created for Famine Early Warning Systems, *Int. J. Remote Sens.,*18: 2147-2159, 1997.

Hessels, T. M.: Comparison and Validation of Several Open Access Remotely Sensed Rainfall Products for the Nile Basin, PhD Thesis, Delft University of Technology, The Netherlands, 2015.

Hirsch, R. M., and Slack, J. R.: A Nonparametric Trend Test for Seasonal Data with Serial Dependence, *Water Resour. Res*, 20, 727-732, 1984.

Hong Wu., SOH, L. K., SAMAL, A., HONG, T., MARX, D., and CHEN, X.: Upstream-Downstream Relationships in Terms of Annual Streamflow Discharges and Drought Events in Nebraska, *JWRP*, 1, 299-315, 2009.

Hsu, K. X. Gao, S. Sorooshian, and H. Gupta.: Precipitation estimation from remotely sensed information using artificial neural networks. *J. Appl.Meteorol., 36*: 1176-1190, 1997.

Hu, Q., D. Yang, Z. Li, A. Mishra Y. Wang, and H. Yang.: Multi-Scale Evaluation of Six High-Resolution Satellite Monthly Rainfall Estimates over a Humid Region in China With Dense Rain Gauges, *Int. J. Remote Sens.*, 35, 1272-1294, 2014.

Huang, M., & Zhang, L.: Hydrological responses to conservation practices in a catchment of the Loess Plateau, China, *Hydrol. Process., 18*(10), 1885-1898, 2004.

Huber, U. M., H. K. Bugmann, and M. A. Reasoner.: Global change and mountain regions: an overview of current knowledge, Springer, Doderecht, Netherlands, 2006.

Huffman, G. D. Bolvin, E. Nelkin, D. Wolff, R. Adler, G. Gu, and E. Stocker.:The TRMM Multisatellite Precipitation Analysis (TMPA): Quasi-Global, Multiyear, Combined-Sensor Precipitation Estimates at Fine Scales, *J. Hydrometeorol.*, 8: 38-55, 2007.

Huffman, G. F. Adler, B. Rudolf, U. Schneider, and P. Keehn.: The Global Precipitation Climatology Project (GPCP) Combined Precipitation Dataset, *BAMS*, 78: 5-20, 1997.

Hughes, D. A.: Modelling semi-arid and arid hydrology and water resources: the southern Africa experience, in: Hydrological Modelling in Arid and Semi-Arid Areas, Wheater, H., Sorooshian, S., and Sharma D., (eds). Cambridge Press, New York, USA, 2008

Hurkmans, R. T. W. L., Terink, W., Uijlenhoet, R., Moors, E. J., Troch, P. A., Verburg, P. H.: Effects of land use changes on streamflow generation in the Rhine basin, *Water Resour. Resear.* 45(6), 2009.

Hurni, H., Tato, K., & Zeleke, G.: The implications of changes in population, land use, and land management for surface runoff in the upper Nile basin area of Ethiopia., *Mount. Res. Devel.* 25(2), 147-154, 2005.

Hurni, K., Zeleke, G., Kassie, M., Tegegne, B., Kassawmar, T., Teferi, E., Moges, A., Tadesse, D., Ahmed, M., Degu, Y., Kebebew, Z., Hodel, E., Amdihun, A., Mekuriaw, A., Debele, B., Deichert, G., Hurni,H.:Economics of Land Degradation (ELD) Ethiopia Case Study. Soil Degradation and Sustainable Land Management in the Rainfed Agricultural Areas of Ethiopia: an Assessment of the Economic Implications, Report for the Economics of Land Degradation Initiative. pp. 94, 2015.

IPCC: Climate Change.: The Physical Science Basis. Contribution of Working Group I to the Fourth Assessment Report of the Intergovernmental Panel on Climate Change, Cambridge University Press, Cambridge, United Kingdon, New York, USA, 2007.

Jaafar, H. H., & Ahmad, F. A.: Time series trends of Landsat-based ET using automated calibration in METRIC and SEBAL: The Bekaa Valley, Lebanon., *Remote Sens. Environ*, In press, 2009.

Jajarmizadeh, M., Harun, S., & Salarpour, M.: A review on theoretical consideration and types of models in hydrology, *J. Environmental Sci. and Technology*, 5(5), 249, 2012.

Jensen, J.:Introductory Digital Image Processing: A Remote Sensing Perspec-tive Prentice Hall, Upper Saddle River, NY, 2005.

Jensen, R.: Introductory Digital Image Processing: A Remote Sensing Perspective (2d ed. ed.). New Jersey: Prentice-Hall, Englewood Cliffs, NY, 1996..

Jiang, S., L. Ren, Y. Hong, B. Yong, X. Yang, F. Yuan.: Comprehensive Evaluation of Multi-Satellite Precipitation Products with a Dense Rain Gauge Network and Optimally Merging their Simulated Hydrological Flows Using the Bayesian Model Averaging Method, *J. Hydrol.*, 452: 213-225, 2012.

Jones, J. R., Schwartz, J. S., Ellis, K. N., Hathaway, J. M., and Jawdy, C. M.: Temporal variability of precipitation in the Upper Tennessee Valley, *J. Hydrol. Regional Studies*, 3, 125-138, 2015.

Joyce, R. J. Janowiak, P. Arkin, and P. Xie.: CMORPH: a Method That Produces Global Precipitation Estimates from Passive Microwave and Infrared Data at High Spatial and Temporal Resolution, *J. Hydrometeorol.*, 5: 487-503, 2004.

Karnieli, A., & Ben-Asher, J.: A daily runoff simulation in semi-arid watersheds based on soil water deficit calculations, *J. Hydrol.*, 149(1–4), 9-25, 1993.

Karssenberg, D., Schmitz, O., Salamon, P., de Jong, K., & Bierkens, M. F.: A software framework for construction of process-based stochastic spatio-temporal models and data assimilation, *Environ. Modell. Softw.* 25, 489-502, 2010.

Katsanos, D., Retalis, A., and Michaelides, S.: Validation of a high-resolution precipitation database (CHIRPS) over Cyprus for a 30-year period, *Atmos. Res.*, 169, 459-464, 2015.

Kebede, S., Travi, Y., Alemayehu, T., & Marc, V.: Water balance of Lake Tana and its sensitivity to fluctuations in rainfall, Blue Nile basin, Ethiopia, *J. Hydrol.*, 316(1), 233-247, 2006.

Kendall, M.: Rank Correlation Methods, Charles Griffi, London, 1975.

Kifle, M., Gebremicael, T. G., Girmay, A., and Gebremedihin, T.: Effect of surge flow and alternate irrigation on the irrigation efficiency and water productivity of onion in the semi-arid areas of North Ethiopia, *Agri. Water Manage*, 187, 69-76, 2017.

Kifle, M., Gebretsadikan, T.G.: Yield and water use efficiency of furrow irrigated potato under regulated deficit irrigation, Atsibi-Wemberta, North Ethiopia, *Agri. Water Manage*, 170, 133-139, 2016.

Kifle, W.: Creating new groundwater through landscape restoration in Tigray, northern Ethiopia: an approach for climate change., International Conference on Improving Food Security in the Face of Climate Change in Africa, July 13-15, 2015, Mekelle, Ethiopia, 2015.

Kim, U., Kaluarachchi, J. J., and Smakhtin, V. U.: Climate change impacts on hydrology and water resources of the Upper Blue Nile River Basin, Ethiopia, IWMI Research Report., 126, 27 pp., 2008.

Kim, U., Kaluarachchi, J. J., and Smakhtin, V.: Climate change impacts on hydrology and water resources of the Upper Blue Nile River Basin, Ethiopia, International Water Management Institute Research Report. 126, 27 p, 2008.

Kimani, M., C. B., Hoedjes, and S. Zhongbo.: An Assessment of Satellite-Derived Rainfall Products Relative to Ground Observations over East Africa, *Remote Sens.*, 9: 430, 2017.

King, J., Brown, C., & Sabet, H.: A scenario-based holistic approach to environmental flow assessments for rivers, *River. Res. Appl.*, 19(5-6), 619-639, 2003.

King, R. S., Baker, M. E., Whigham, D. F., Weller, D. E., Jordan, T. E., Kazyak, P. F., & Hurd, M. K.: Spatial considerations for linking watershed land cover to ecological indicators in streams, *Eco. Appl. 15*(1), 137-153, 2005.

Kiptala, J. K., Mohamed, Y., Mul, M. L., & Van Der Zaag, P.: Mapping evapotranspiration trends using MODIS and SEBAL model in a data scarce and heterogeneous landscape in Eastern Africa, *Water Res. Resear.*, 49(12), 8495-8510, 2013b.

Kiptala, J. K., Mohamed, Y., Mul, M. L., Cheema, M. J. M., & Van der Zaag, P.: Land use and land cover classification using phenological variability from MODIS vegetation in the Upper Pangani River Basin, Eastern Africa, *Phys. Chem.Earth. 66*, 112-122, 2013a.

Kiptala, J., Mul, M., Mohamed, Y.,Van der Zaag, P.: Modelling stream flow and quantifying blue water using a modified STREAM model for a heterogeneous, highly utilized and data-scarce river basin in Africa, *Hydrol. Earth Syst. Sci.* 18, 2014.

Kiros, G., A. Shetty, and Nandagiri, L.: Analysis of Variability and Trends in Rainfall over Northern Ethiopia, *Arab. J. Geosci.*, 9: 451, 2016.

Knapp, K., S. Ansari, C. Bain, M. Bourassa, M. Dickinson, C. Funk, and. Huffman, G.: Globally Gridded Satellite Observations for Climate Studies, *BAMS.*, 92: 893-907, 2011.

Köhler, L., Mulligan, M., Schellekens, J., Schmid, S., Tobón, C.: Hydrological impacts of converting tropical montane cloud forest to pasture, with initial reference to northern Costa Rica, Final Technical Report DFID-FRP Project no. R799, 2006.

Kopačková, V., Chevrel, S., Bourguignon, A., & Rojík, P.: Application of high altitude and ground-based spectroradiometry to mapping hazardous low-pH material derived from the Sokolov open-pit mine, *J. Maps. 8*(3), 220-230. 2012.

Kralovec, V., Kliment, Z., & Matouskova, M.: Evaluation of runoff response on the basis of a comparative paired research in mountain catchments with the different land use: case study of the Blanice River, Czechia. *Geografie, 121*(2), 209-234, 2016.

Laaha, G., & Blöschl, G.: A comparison of low flow regionalisation methods catchment grouping, *J. Hydrol.*, 323(1), 193-214, 2006.

Lacombe, G., Cappelaere, B., & Leduc, C.: Hydrological impact of water and soil conservation works in the Merguellil catchment of central Tunisia, *J. Hydrol.*, 359(3), 210-224, 2008.

Li, H., and Sivapalan, M.: Effect of spatial heterogeneity of runoff generation mechanisms on the scaling behaviour of event runoff responses in a natural river basin, *Water Resour. Res.* 47, 2011

Li, L. J., Zhang, L., Wang, H., Wang, J., Yang, J. W., Jiang, D. J. Qin, D. Y.: Assessing the impact of climate variability and human activities on streamflow from the Wuding River basin in China, *Hydrol. Process.*, *21*(25), 3485-3491, 2007.

Liu, B., De Smedt, F., Hoffmann, L., & Pfister, L.: Assessing land use impacts on flood processes in complex terrain by using GIS and modeling approach, *Environ. Model. Asses.* *9*(4), 227-235, 2005.

Liu, Y., Wang, L., & Long, H.: Spatio-temporal analysis of land-use conversion in the eastern coastal China during 1996–2005, *J. Geog. Sci. 18*(3), 274-282, 2008.

Longobari, A. and Villani, P.: Trend analyses of annual and seasonal rainfall time series in the Mediterranean area, *Int. J. Climatol.*, 30, 1538-1546, 2009.

López, P., Wanders, N., Schellekens, J., Renzullo, L., Sutanudjaja, E., Bierkens, M.: Improved large-scale hydrological modelling through the assimilation of streamflow and downscaled satellite soil moisture observations, *Hydrol. Earth Syst. Sci.* 20, 3059-307, 2016.

Love, D., Uhlenbrook, S., Corzo-Perez, G., Twomlow, S., and van der Zaag, P.: Rainfall–interception–evaporation–runoff relationships in a semi-arid catchment, northern Limpopo basin, Zimbabwe, *Hydrol. Sci. J.*, 55, 687-703, 2010.

Lu, Z., Zou, S., Qin, Z., Yang, Y., Xiao, H., Wei, Y., Xie, J.: Hydrologic Responses to Land Use Change in the Loess Plateau: Case Study in the Upper Fenhe River Watershed, *Adv. Meteorol.* 10, 2015.

Ma, Z., Kang, S., Zhang, L., Tong, L., and Su, X.: Impacts of climate variability and human activity on streamflow for a river basin in arid region of northwest China, *J. Hydrol.*, 352, 239-249, 2008.

Martinez-Mena, M., Albaladejo, J., & Castillo, V.: Factors influencing surface runo generation in a Mediterranean semi-arid environment: Chicamo watershed, SE Spain, *Hydrol. Process, 12*, 741-754, 2008.

Masih, I., Uhlenbrook, S., Maskey, S., and Smakhtin, V.: Streamflow trends and climate linkages in the Zagros Mountains, Iran, *Climatic Change*, 104, 317-338, 2011.

Mathews, R. and Richter, B. D.: Application of the Indicators of Hydrologic Alteration Software in Environmental Flow Set-ting1, *JAWRA J. Am. Water Resour. Assoc.*, 43, 1400–1413, 2007.

Medvigy, D., and Beaulieu, C.: Trends in daily solar radiation and precipitation coefficients of variation since 1984, *J. Climate*, 25, 1330-1339, 2012.

Meng, J., L. Li, Z. Hao, J. Wang, and Shao, Q.: Suitability of TRMM Satellite Rainfall in Driving a Distributed Hydrological Model in the Source Region of Yellow River., *J. Hydrol.*, 509: 320-332.

Mesfin, H, Gebremedhin, A, Gebremicael, T.G.: Hydrological response under climate change scenario using SWAT Model: The case of Ilala watershed, Northern Ethiopia, *Modeling Earth Systems and Environment*, 4(1), 437-447, 2018.

Mohamed, Y., and Savenije, H.: Impact of climate variability on the hydrology of the Sudd wetland: signals derived from long term (1900–2000) water balance computations, *Wetl. Ecol.Manage*, 191-198, 2014.

Mohamed, Y.A., B.J.J.M. van den Hurk, H.H.G. Savenije, and W.G.M. Bastiaanssen, Hydro-climatology of the Nile: results from a regional climate model, *Hydrol. Earth Sys.Sci.* 2: 263–278.,2005.

Moreno, J. F., C. M. Mannaerts, and V. Jetten.: Influence of Topography on Rainfall Variability in Santiago Island, Cape Verde, *Int. J. Climatol.*, 34: 1081-1097, 2014.

Moriasi, D. N., Arnold, J. G., Van Liew, M. W., Bingner, R. L., Harmel, R. D., & Veith, T. L.: Model evaluation guidelines for systematic quantification of accuracy in watershed simulations, *T. ASABE.* 50, 885-900, 2007.

Morin, J., & Benyamini, Y. (1977). Rainfall infiltration into bare soils, *Water Res. Resear.* 13(5), 813-817, 1997.

Mu, X., Zhang, L., McVicar, T. R., Chille, B., & Gau, P.: Analysis of the impact of conservation measures on stream flow regime in catchments of the Loess Plateau, China, *Hydrol. Process.,* 21(16), 2124-2134, 2007.

Mu, X., Zhang, L., McVicar, T. R., Chille, B., & Gau, P.: Analysis of the impact of conservation measures on stream flow regime in catchments of the Loess Plateau, China, *Hydrol.Process.,* 21(16), 2124-2134, 2007.

Munro, R. N., Deckers, J., Haile, M., Grove, A., Poesen, J., & Nyssen, J.: Soil landscapes, land cover change and erosion features of the Central Plateau region of Tigrai, Ethiopia: Photo-monitoring with an interval of 30 years, *Catena,* 75, 55-64, 2008.

Návar, J.: Hydro-climatic variability and perturbations in Mexico's north-western temperate forests, *Ecohydrol.,* 8, 1065-1072, 2015.

Negusse, T., Yazew, E., and Tadesse, N.: Quantification of the impact of integrated soil and water conservation measures on groundwater availability in Mendae catchment, Abraha We-Atsebaha, eastern Tigray, Ethiopia, *MEJS.,* 5, 117-136, 2013.

Nepal, S., Flügel, W.-A., & Shrestha, A. B.: Upstream-downstream linkages of hydrological processes in the Himalayan region, *Ecol. Process.* 3 (1), 1-16, 2014.

NMSA (National Meteorological service Agency): Climatic and agro-climatic resources of Ethiopia. NMSA Meteorological Research Report Series. V1, No. 1, Addis Ababa, 137p, 1996.

Nyssen, J., Clymans, W., Descheemaeker, K., Poesen, J., Vandecasteele, I., Vanmaercke, M., Zenebe, A., Van Camp, M., Haile, M., and Haregeweyn, N.: Impact of soil and water conservation measures on catchment hydrological response a case in north Ethiopia, *Hydrol. Process,* 1880-1895, 2010.

Nyssen, J., Frankl, A., Haile, M., Hurni, H., Descheemaeker, K., Crummey, D., . . . Moeyersons, J.: Environmental conditions and human drivers for changes to north Ethiopian mountain landscapes over 145 years, *Sci. Total Environ.,* 485, 164-179., 2014.

Nyssen, J., Frankl, A., Zenebe, A., Deckers, J., & Poesen, J.: Land management in the northern Ethiopian highlands: local and global perspectives; past, present and future. *Land Degrad. Develop.,* 2015a.

Nyssen, J., H. Vandenreyken, J. Poesen, J. Moeyersons, J. Deckers, M. Haile, and G. Govers.: Rainfall Erosivity and Variability in the Northern Ethiopian Highlands, *J. Hydrol.,* 311: 172-187, 2005.

Nyssen, J., Haile, M., Moeyersons, J., Poesen, J., & Deckers, J.: Soil and water conservation in Tigray (Northern Ethiopia): the traditional dagat technique and its integration with introduced techniques, *Land deg. Develop.,* 11, 199-208, 2000.

Nyssen, J., Poesen, J., Moeyersons, J., Deckers, J., Haile, M., & Lang, A.: Human impact on the environment in the Ethiopian and Eritrean highlands a state of the art, *Earth.Sci. Rev.,* 64(3–4), 273-320, 2004.

Nyssen,J., Poesen, J., Lanckriet, S., Jacob, M., Moeyersons, J., Haile, M., Haregeweyn, N., Munro, N., Descheemaeker, K., Adgo, E., Frankl, A., Deckers, J.: Land degradation in the Ethiopian highlands. In Landscapes and landforms of Ethiopia, Billi,P. (ed). Springer: Dordrecht, The Netherlands, 369–385, 2015b.

Oliver, M. A., Webster, R.: Kriging: a method of interpolation for geographical information systems, *Int. J. Geogr. Inf. Sci.* 4, 313-332, 1990.

Ott, B., Uhlenbrook, S.: Quantifying the impact of land-use changes at the event and seasonal time scale using a process-oriented catchment model, *Hydrol.Earth Syst. Sci.* 8, 62-78, 2004.

Ouma, Y. O., T. Owiti, E. Kipkorir, J. Kibiiy, and R. Tateishi.: Multitemporal Comparative Analysis of TRMM-3B42 Satellite-Estimated Rainfall with Surface Gauge Data at Basin Scales: Daily, Decadal and Monthly Evaluations, *Int.J. Remote Sens.,* 33: 7662-7684, 2012.

Pano, N., Frasheri, A., and Avdyli, B.: The climatic change impact in water potential processes on the Albanian hydrographic river network, Proceedings of the 5th Biennial Conference of the International Environmental Modelling and Software Society, *iEMSs* 2010, 927-933, 2010.

Pettittt, A. N.: A Non-Parametric Approach to the Change-Point Problem, *Journal of the Royal Statistical Society, Series C (Applied Statistics)*, 28, 126-135, 1979.

Pilgrim, D., Chapman, T., & Doran, D.: Problems of rainfall-runoff modelling in arid and semiarid regions, *Hydrol. Sci. J., 33*(4), 379-400, 1998.

Poff, N. L., & Zimmerman, J. K.: Ecological responses to altered flow regimes: a literature review to inform the science and management of environmental flows, *Freshwater Biol., 55*(1), 194-205, 2010.

Pontius, R. G., Shusas, E., & Mceachern, M.: Detecting important categorical land changes while accounting for persistence, *Agric. Ecosyst. Environ. 101*, 251-268, 2004.

Pushpalatha, R., Perrin, C., Moine, N. L., & Andréassian, V.: A review of efficiency criteria suitable for evaluating low-flow simulations, *J. Hydrol., 420-421*, 171-182, 2012.

Reed, S., Koren, V., Smith, M., Zhang, Z., Moreda, F., Seo, D.-J., & Dmip Participants, a.: Overall distributed model intercomparison project results, *J. Hydrol., 298*(1–4), 27-60, 2004.

Refsgaard, J. C., & Knudsen, J.: Operational validation and intercomparison of different types of hydrological models, *Water Res. Resear., 32*(7), 2189-2202, 1996..

Richter, R., & Schlapfer, D.: Atmospheric/topographic correction for airborne imagery: ATCOR-4 User Guide. *DLR IB*, 565-502, 2012.

Sanyal, J., Densmore, A. L., & Carbonneau, P.: Analysing the effect of land-use/cover changes at sub-catchment levels on downstream flood peaks: A semi-distributed modelling approach with sparse data, *CATENA, 118*(0), 28-40, 2014.

Sapiano, M. R. P., and P. A Arki. 2009. "An Intercomparison and Validation of High-Resolution Satellite Precipitation Estimates with 3-Hourly Gauge Data, *J. Hydrometeorol.,* 10: 149-166.

Saraiva Okello, A. M. L., Masih, I., Uhlenbrook, S., Jewitt, G. P. W., van der Zaag, P., and Riddell, E.: Drivers of spatial and temporal variability of streamflow in the Incomati River basin, *Hydrol. Earth Syst. Sci.*, 19, 657-673, 2015.

Savenije, H. H., Hoekstra, A. Y., Van der Zaag, P.: Evolving water science in the Anthropocene, *Hydrol. Earth Syst.Sci.* 18, 319-332, 2014.

Savenije, H.: HESS Opinions "Topography driven conceptual modelling (FLEX-Topo)", *Hydrol. Earth Syst. Sci.*, 14(12), 2681-2692, 2010.

Savenije, H.: HESS Opinions" The art of hydrology", *Hydro. Earth Sys.Sci.,* 13(2), 157-161, 2009.

Schellekens, J.: Wflow, a flexible hydrological model. OpenStream wflow documentation release 1.0 RC1, Deltares, Delft, The Netherlands, 2014.

Schmidt, E., & Zemadim, B.: Hydrological modelling of sustainable land management interventions in the Mizewa watershed of the Blue Nile basin, ILRI report, Adis Abeab, Ethiopia, 2013.

Seibert, J., McDonnell, J.: Land-cover impacts on streamflow: a change-detection modelling approach that incorporates parameter uncertainty, *Hydrol. Sci. J.*, 55, 316-332, 2010.

Seleshi, Y., and Zanke, U.: Recent changes in rainfall and rainy days in Ethiopia, *Int. J. Climatol.*, 24, 973-983, 2004.

Shapiro, S. S., Wilk, M. B.: An analysis of variance test for normality (complete samples). *Biometrika.* 52, 591-611, 1965.

Shi, W., Yu, X., Liao, W., Wang, Y., & Jia, B.: Spatial and temporal variability of daily precipitation concentration in the Lancang River basin, China, *J. Hydrol.* 495(0), 197-207, 2013.

Shukla, S., McNally, A., Husak, G., Funk, C.: A seasonal agricultural drought forecast system for food-insecure regions of East Africa, *Hydrol. Earth Syst. Sci.* 18, 3907–921, 2014.

Silveira, L., & Alonso, J.: Runoff modifications due to the conversion of natural grasslands to forests in a large basin in Uruguay, *Hydrol. Process.*, 23(2), 320-329, 2009.

Simpson, J., Adler, R. F., and North, G. R.: A proposed tropical rainfall measuring mission (TRMM) satellite, *BAMS.*, 69, 278-295, 1988.

Sivapalan, M., Blöschl, G., Zhang, L., & Vertessy, R.: Downward approach to hydrological prediction, *Hydrol. Process.*, *17*, 2101-2111, 2003..

Sivapalan, M., Blöschl, G., Zhang, L.,Vertessy, R., 2003. Downward approach to hydrological prediction, *Hydrol. Process.* 17, 2101-2111, 2003.

Smakhtin, V. U.: Low flow hydrology: a review, *J. Hydrol.*, 240(3), 147-186, 2001.

Smit, H., Muche, R., Ahlers, R., van der Zaag, P.: The Political Morphology of Drainage—How Gully Formation Links to State Formation in the Choke Mountains of Ethiopia. *World Development*, 98, 231-244, 2017.

Ssegane, H., Amatya, D. M., Chescheir, G. M., Skaggs, W. R., Tollner, E. W., & Nettles, J. E.: Consistency of hydrologic relationships of a paired watershed approach, *Am. J. Clim. Change*, 2 (2), 147-164, 2013.

Steenhuis, T. S., Collick, A. S., Easton, Z. M., Leggesse, E. S., Bayabil, H. K., White, E. D., Ahmed, A. A.: Predicting discharge and sediment for the Abay (Blue Nile) with a simple model, *Hydrol. Process.*, 23(26), 3728-3737, 2009.

Suliman, A. H. A., Jajarmizadeh, M., Harun, S., & Darus, I. Z. M.: Comparison of Semi-Distributed, GIS-Based Hydrological Models for the Prediction of Streamflow in a Large Catchment, *Water Res. Manag.*, 29, 3095-3110, 2015.

Sunilkumar, K., R. T. Narayana, K. Saikranthi, and R M.: Purnachandra.: Comprehensive Evaluation of Multisatellite Precipitation Estimates over India Using Gridded Rainfall Data, *J. Geophy. Resear-Atmos.*, 120: 8987-9005, 2015.

Sushant, S., Balasubramani, K., and Kumaraswamy, K.: Spatio-temporal Analyses of Rainfall Distribution and Variability in the Twentieth Century, Over the Cauvery Basin, South India, in: Environmental Management of River Basin Ecosystems, 21-41, 2015.

Tadesse, N., Bairu, A., & Bheemalingeswara, K.: Suitability of Groundwater Quality for Irrigation with Reference to Hand Dug Wells, Hantebet Catchment, Tigray, Northern Ethiopia, *MEJS.*, 3, 31-47, 2011.

Taniguchi, M., 2012. Subsurface hydrological responses to land cover and land use changes: Springer Science & Business Media. Utrecht, The Netherlands.

Taye, G., Poesen, J., Deckers, J., Tekka, D., Haregeweyn, N., van Wesemael, B., & Nyssen, J.: The effect of soil and water conservation treatments on rainfall-runoff response and soil losses in the Northern Ethiopian Highlands: the case of May Leiba catchment. In J. Nyssen, Asfawesen, A., Francesco, D, Mohamed, U. (Ed.), Excursion guide: post-conference excursion: geomorphological hazards, land degradation and resilience in the North Ethiopian highlands (pp. 86-92). Addis Abeba, Ethiopia, 2011.

Taye, G., Poesen, J., Vanmaercke, M., van Wesemael, B., Martens, L., Teka, D., . . . Hallet, V.: Evolution of the effectiveness of stone bunds and trenches in reducing runoff and soil loss in the semi-arid Ethiopian highlands, *Zeitschrift für Geomorphologie, 59*(4), 477-493, 2015.

Taye, M.T., and Willems, P.: Identifying Sources Of Temporal Variability In Hydrological Extremes Of The Upper Blue Nile Basin, *J. Hydrol.*, 499: 61-70, 2013.

Taylor, V., Schulze, R., and Jewitt, G. P. W.: Application of the Indicators of Hydrological Alteration method to the Mkomazi River, KwaZulu-Natal, South Africa, *Afr. J. Aquatic Sci.*, 28, 1–11, 2003.

Tefera, W.: Integtating Geo-information for the management of river basins in Ethiopian part of the Nile basin, MSc Thesis, ITC, Enschede, The Netherlands, 2003.

Teferi, E., Bewket, W., Uhlenbrook, S., & Wenninger, J.: Understanding recent land use and land cover dynamics in the source region of the Upper Blue Nile, Ethiopia: Spatially explicit statistical modeling of systematic transitions.,*Agric. Ecosyst. Environ. 165*, 98-117, 2013.

Tekle, K., & Hedlund, L.: Land Cover Changes Between 1958 and 1986 in Kalu District, Southern Wello, Ethiopia, *Mount. Res. Dev.* 20(1), 42-51, 2000.

Tekleab, S., Mohamed, Y., and Uhlenbrook, S.: Hydro-climatic trends in the Abay/Upper Blue Nile basin, Ethiopia, *Phys. Chem. Earth, Parts A/B/C*, 61-62, 32-42, 2013.

Tekleab, S., Mohamed, Y., Uhlenbrook, S., Wenninger, J.: Hydrologic responses to land cover change: the case of Jedeb mesoscale catchment, Abay/Upper Blue Nile basin, Ethiopia, *Hydrolo. Process.* 28, 5149-5161, 2014.

Tenenhaus, M.: La régression PLS, théorie et pratique. Editions Technip, Paris, France, 1998.

Tesemma, Z. K., Mohamed, Y. A., and Steenhuis, T. S.: Trends in rainfall and runoff in the Blue Nile Basin: 1964–2003, *Hydrol. Process*, 24, 3747-3758, 2010.

Tesfaye, A., Negatu, W., Brouwer, R., Van der Zaag, P., 2014. Understanding soil conservation decision of farmers in the Gedeb watershed, Ethiopia. *Land Degradation & Development* 25, 71-79

Tesfaye, S., Birhane, E., Leijnse, T., van der Zee, S.E.A.T.M.: Climatic controls of ecohydrological responses in the highlands of northern Ethiopia, *Sci.Total Environ.*, 609, 77-91, 2017.

Thiemig, V., Rojas, R., Zambrano-Bigiarini, M., Levizzani, V., and De Roo, A.: Validation of satellite-based precipitation products over sparsely gauged African river basins, *J. Hydrometeorol.*, 13, 1760-1783, 2012.

Toté, C., D. Patricio, H. Boogaard, R.van der Wijngaart, E.Tarnavsky, and C. Funk.: "Evaluation of Satellite Rainfall Estimates for Drought and Flood Monitoring in Mozambique, *Remote Sens.* 7: 1758, 2015.

Tsegaye, D., Moe, S., Paul, V., & Aynekul, E.: Land-use/cover dynamics in Northern Afar rangelands, Ethiopia, *Agric. Ecosyst. Environ.* 139, 174-180, 2010.

Uhlenbrook, S., Roser, S., Tilch, N.: Hydrological process representation at the meso-scale: the potential of a distributed, conceptual catchment model. *J. Hydrol.* 291, 278-296, 2004.

Uhlenbrook, S., Seibert, J., Leibundgut, C., & Rodhe, A.: Prediction uncertainty of conceptual rainfall-runoff models caused by problems in identifying model parameters and structure, *Hydrol.Sci. J.*, 779-797.

UNESCO-IHE : pySEBAL, Delft, The Netherlands

van de Giesen, N., Stomph, T. J., & de Ridder, N.: Surface runoff scale effects in West African watersheds: modeling and management options, *Agri. Water Manag.*, 72(2), 109-130, 2005.

Van der Ent, R. J., H.H. Savenije, B. Schaefli, and S.C. Steele-Dunne.: Origin and Fate of Atmospheric Moisture over Continents, *Water Res. Resear.*, 46: W09525, 2010.

van Eekelen, M. W., Bastiaanssen, W. G. M., Jarmain, C., Jackson, B., Ferreira, F., van der Zaag, P., Luxemburg, W. M. J.: A novel approach to estimate direct and indirect water withdrawals from satellite measurements: A case study from the Incomati basin, *Agri. Ecosy. Environ.*, 200(0), 126-142, 2015.

Van Vliet, J., Bregt, A. K., & Hagen-Zanker, A.: Revisiting Kappa to account for change in the accuracy assessment of land-use change models, *Ecolo. Model. 222*, 1367-1375, 2010.

Verma, D.: A physical model of the rainfall–runoff relationship for semiarid lands. Canberra Symposium on the hydrology of areas of low precipitation, IAHS Publication, 128, 215–222, 1979.

Vertessy, R. A., & Elsenbeer, H.: Distributed modeling of storm flow generation in an Amazonian rain forest catchment: Effects of model parameterization. *Water Res. Resear.*, 35(7), 2173-2187, 1999.

Vertessy, R. A., Elsenbeer, H.: Distributed modeling of storm flow generation in an Amazonian rain forest catchment: Effects of model parameterization. *Water Resour. Resear.* 35, 2173-2187, 1999.

Virgo, K.J., Munro, R.N.: Soil and erosion features of the Central Plateau region of Tigray, Ethiopia, *Geoderma* 20 (2), 131–157, 1977.

Viste, E., and A. Sorteberg.: Moisture Transport into the Ethiopian Highlands. *Int. J. Climatol.*, 33: 249-263, 2012.

Wafi A.: Historical Land Use/Land Cover Classification Using Remote Sensing. A Case Study of the Euphrates River Basin in Syria: Springer, 2013.

Wagener, T.: Can we model the hydrological impacts of environmental change? *Hydrol. Process., 21*, 3233-3236, 1978, 2007.

Walraevens, K., Gebreyohannes Tewolde, T., Amare, K., Hussein, A., Berhane, G., Baert, R., Ronsse, S., Kebede, S., Van Hulle, L., Deckers, J., Martens, K.: Water balance components for sustainability assessment of groundwater-dependent agriculture: example of the mendae plain (Tigray, Ethiopia), *Land Degrad. Dev.*, 26 (7), 725–736, 2051.

Wang, G., Yu, M., & Xue, Y.:. Modeling the potential contribution of land cover changes to the late twentieth century Sahel drought using a regional climate model: impact of lateral boundary conditions, *Clim. Dynam.*, 1-21, 2015.

Wang, H., Gao, J.-e., Zhang, S.-l., Zhang, M.-j., & Li, X.-h.: Modeling the Impact of Soil and Water Conservation on Surface and Ground Water Based on the SCS and Visual Modflow, *PLoS ONE*, 8(11), e79101, 2013.

Wang, S., liu, S., & Ma, T.: Dynamics and changes in spatial patterns of land use in Yellow River Basin, China., *Land Use Policy*, 27, 313–323, 2013.

Wang, W., L. Hui, Z. Tianjie, J. Lingmei, and S. Jianchen.: Evaluation and comparison of newest GPM and TRMM products over Mekong River Basin at daily scale." Geoscience and Remote Sensing Symposium (IGARSS), 2016 IEEE International. IEEE 10: 2540-2549, 2017.

Wang, W., Shao, Q., Yang, T., Peng, S., Xing, W., Sun, F., and Luo, Y.: Quantitative assessment of the impact of climate variability and human activities on runoff changes: a case study in four catchments of the Haihe River basin, China, *Hydrol. Process*, 27, 1158-1174, 2013.

Wang, W., Wei, J., Shao, Q., Xing, W., Yong, B., Yu, Z., and Jiao, X.: Spatial and temporal variations in hydro-climatic variables and runoff in response to climate change in the Luanhe River basin, China, Stoch, *Env. Res. Risk. A.*, 29, 1117-1133, 2015.

Wang, X., Zhang, J., & Babovic, V.: Improving real-time forecasting of water quality indicators with combination of process-based models and data assimilation technique, *Ecol. Indic.*, 66, 428-439, 2016.

Wang, Y., Liu, Y., & Jin, J.: Contrast Effects of Vegetation Cover Change on Evapotranspiration during a Revegetation Period in the Poyang Lake Basin, China. *Forests.* 9(4) 217, 2018.

Watson, F., Vertessy, R., McMahon, T., Rhodes, B., & Watson, I.: Improved methods to assess water yield changes from paired-catchment studies: application to the Maroondah catchments, *Forest Ecol. Manag.*, 143(1–3), 189-204, 2001.

Wheater, H., Sorooshian, S., & Sharma, K. D.: Hydrological modelling in arid and semi-arid areas: Cambridge University Press, Cambridge, UK, 2007.

White, E. D., Easton, Z. M., Fuka, D. R., Collick, A. S., Adgo, E., McCartney, M., Steenhuis, T. S.: Development and application of a physically based landscape water balance in the SWAT model, *Hydrol. Process.*, 25(6), 915-925, 2011.

Wilk, J., and Hughes, D. A.: Simulating the impacts of land-use and climate change on water resource availability for a large south Indian catchment, *Hydrol. Sci. J.*, 47, 19-30, 2002.

Wittenberg, H.: Effects of season and man-made changes on baseflow and flow recession: case studies, *Hydrol. Process.* 17(11), 2113-2123, 2003.

WMO: World Meteorological Organization Guide to Hydrological Practices: Data Acquisition and Processing, Analysis, Forecasting and Other Applications." Geneva, Switzerland, 1994.

Woldearegay K, & Tamene L.: Water Harvesting and Irrigation Development: Enhancing Productivity under Rainfall Variability. Paper presented at the Training of Trainers (ToT) Workshop, 19-21 April 2017, Michew, Ethiopia.

Woldearegay, K., Tamene, L., Mekonnen, K., Kizito, F., & Bossio, D.: Fostering Food Security and Climate Resilience Through Integrated Landscape Restoration Practices and Rainwater Harvesting/Management in Arid and Semi-arid Areas of Ethiopia Rainwater-Smart Agriculture in Arid and Semi-Arid Areas, (pp. 37-57): Springer, 2018.

Woldesenbet, T. A., Elagib, N. A., Ribbe, L., & Heinrich, J.: Hydrological responses to land use/cover changes in the source region of the Upper Blue Nile Basin, Ethiopia, *Sci. Total Environ*, 575, 724-741, 2017.

Wondie, M., Schneider, W., Melesse, A., & Teketay, D.: Spatial and Temporal Land Cover Changes in the Simen Mountains National Park, a World Heritage Site in Northwestern Ethiopia., *Remote Sens.* 3, 752-766, 20111.

Worku, A., Adugna, B., Wubeshet, H., Sisay, S., Mebrahtu, T., Gebreegziabhe, T., & Hailu, Z.: GIZ Ethiopia:Lessons and Experiences in Sustainable Land Management. Addis Abeba, Ethiopia: Deutsche Gesellschaft für Internationale Zusammenarbeit (GIZ) GmbH, 2015.

Worqlul, A. W., Ayana, E. K., Yen, H., Jeong, J., MacAlister, C., Taylor, R., Steenhuis, T. S.: Evaluating hydrologic responses to soil characteristics using SWAT model in a paired-watersheds in the Upper Blue Nile Basin, *Catena*, 163, 332-341, 2018.

Worqlul, A., B. Maathuis, A. A. Adem, S. S. Demissie, S. Langan, and T. S. Steenhuis.: Comparison of Rainfall Estimations by TRMM 3B42, MPEG and CFSR with Ground-Observed Data for the Lake Tana Basin in Ethiopia, *Hydrol.Earth Sys. Sci.*, 18: 4871-4881, 2014.

Xie, P., and A. Arkin.: An Intercomparison of Gauge Observations and Satellite Estimates of Monthly Precipitation, *J. App. Meteorol.*, 34: 1143-1160, 1995.

Xie, P., and Atkin. P.: Global Precipitation: A 17-Year Monthly Analysis Based on Gauge Observations, Satellite Estimates, and Numerical Model Outputs, *BAMS.*, 78: 2539-2558, 1997.

Xiubin, H., Zhanbin, L., Mingde, H., Keli, T., & Fengli, Z. (2003). Down-scale analysis for water scarcity in response to soil–water conservation on Loess Plateau of China, *Agri. Ecosys. Environ.*, 94(3), 355-361, 2003.

Xu R., T. Fuqiang, L. Yang, H. Hu, H. Lu, and A. Hou.: Ground Validation of GPM IMERG and TRMM 3B42V7 Rainfall Products over Southern Tibetan Plateau Based on a High-Density Rain Gauge Network, *J.Geophy.Resear-Atmos.*, 122: 910-924, 2017.

Xu, X., & Yang, D.: Analysis of catchment evapotranspiration at different scales using bottom-up and top-down approaches, *Front Architect Civ. Eng. China.*, 4(1), 65-77, 2010.

Xue, X., Y. Hong, A. S. Limaye, J. J. Gourley, G. J. Huffman, S. I. Khan, and S. Chen.: Statistical and Hydrological Evaluation of TRMM-Based Multi-Satellite Precipitation Analysis over the

Wangchu Basin of Bhutan: Are the Latest Satellite Precipitation Products 3B42V7 Ready for Use in Ungauged Basins, *J. Hydrol.*, 499: 91-99, 2013.

Yair, A., and Kossovsky, A.: Climate and surface properties: hydrological response of small arid and semi-arid watersheds, *Geomorphology*, 42, 43-57, 2002

Yan, R., Gao, J., Li, L.: Modeling the hydrological effects of climate and land use/cover changes in Chinese lowland polder using an improved WALRUS model, *Hydrol. Res.* 47(S1), 84-101, 2016.

Yang, D., Herath, S., & Musiake, K.:. Comparison of different distributed hydrological models for characterization of catchment spatial variability, *Hydrol. Process., 14*(3), 403-416, 2010.

Yazew, H., E.: Development and management of irrigated lands in Tigray, Ethiopia, PhD, UNESCO-IHE, Institute for Water Education, Delft, The Netherlands, 2005.

Yosef, B. A., & Asmamaw, D. K.:. Rainwater harvesting: An option for dry land agriculture in arid and semi-arid Ethiopia. *Int. J. Water Res. Environ. Eng.*, 7(2), 17-28, 2015.

Yu, Y., Zhang, H., Singh, V.P.: Forward prediction of runoff data in data-scarce basins with an improved ensemble empirical mode decomposition (EEMD) model, *Water* 10, 388, 2018.

Yuan, Y., Li, B., Gao, X., Liu, H., Xu, L., & Zhou, C.: A method of characterizing land-cover swap changes in the arid zone of China, *Front. Earth Sci.* 10(1), 74-86, 2016.

Yue, S., Pilon, P., and Phinney, B.: Canadian streamflow trend detection: impacts of serial and cross-correlation, *Hydrol. Sci. J.*, 48, 51-63, 2003.

Yue, S., Pilon, P., Phinney, B., and Cavadias, G.: The influence of autocorrelation on the ability to detect trend in hydrological series, *Hydrol. Process*, 16, 1807-1829, 2002.

Zeleke, G., & Hurni, H. (2001). Implications of Land Use and Land Cover Dynamics for Mountain Resource Degradation in the Northwestern Ethiopia, *Mount. Res. Devel.* 21, 184-191, 2001.

Zenebe, A., Vanmaercke, M., Poesen, J., Verstraeten, G., Haregeweyn, N., Haile, M., Nyssen, J.: Spatial and temporal variability of river flows in the degraded semi-arid tropical mountains of northern Ethiopia. *Z. Geomorphol.*, 57, 143-169, 2013.

Zenebe, A.: Assessment of spatial and temporal variability of river discharge, sediment yield and sediment-fixed nutrient export in Geba River catchment, northern Ethiopia, PhD, Stoch. Env. Res. Risk. A.,, KU Leuven,, Leuven, Belgium, 2009.

Zhan, C. S., Jiang, S. S., Sun, F. B., Jia, Y. W., Niu, C. W., and Yue, W. F.: Quantitative contribution of climate change and human activities to runoff changes in the Wei River basin, China, *Hydrol. Earth Syst. Sci.*, 18, 3069-3077, 2014.

Zhang, Q., Singh, V. P., Sun, P., Chen, X., Zhang, Z., and Li, J.: Precipitation and streamflow changes in China: changing patterns, causes and implications, *J. Hydrol.*, 410, 204-216, 2011.

Zhang, S., Lu, X. X., Higgitt, D. L., Chen, C.-T. A., Han, J., and Sun, H.: Recent changes of water discharge and sediment load in the Zhujiang (Pearl River) Basin, China, *Glob. Planet. Chang*, 60, 365-380, 2008.

Zhao, J., Huang, Q., Chang, J., Liu, D., Huang, S., and Shi, X.: Analyses of temporal and spatial trends of hydro-climatic variables in the Wei River Basin, *Environ. Res.*, 139, 55-64, 2015.

Zhao, M., Zeng, C., Liu, Z., & Wang, S.: Effect of different land use/land cover on karst hydrogeochemistry: A paired catchment study of Chenqi and Dengzhanhe, Puding, Guizhou, SW China, *J. Hydrol.*, 388(1), 121-130, 2010.

APPENDIX A (CHAPTER 3)

Table A-1: Location, altitude and average annual rainfall of the ground measurement observations (Longitude and Latitude in degrees)

Station name	Longitude	Latitude	Altitude (meter)	Average rainfall (mm/year)
Adi-Arkay	38° 07'12"	13°14'00"	1600	1948
Ambagiorgis	37°44'24"	12°36'00"	3010	1157
Dabat	37°54'36"	12°54'36"	2600	838
Debark	38°22'4"	13°41'20"	3003	1273
Debre Tabor	38°9'36"	12°04'40"	2700	1628
Abi-Adi	39°1'12"	13°22'12"	1950	1114
Adigrat	39°12'00"	14°15'00"	2400	512
Adigudom	39°31'12"	13°09'36"	2100	453
Adishehu	39°31'12"	12°56'24"	2500	546
Adwa	38°54'00"	14°42'00"	1960	766
Dengolet	39°00'00"	13°06'00"	2150	692
Edaga Hamus	39°19'48"	14°07'12"	2700	546
Hagere Selam	39°16'12"	13°21'00"	2650	660
Maichew	39°31'48"	12°47'24"	2450	720
Maykental	39°00'00"	13°32'24"	1800	578
Mekele (AP)	39°15'36"	13°09'00"	2200	530
Senkata	39°21'36"	13°34'12"	2500	540
Shire	38° 14' 24"	14°00'36"	1920	1009
Wukuro	39° 21' 36"	13° 27' 36"	1970	600
Korem	39° 26' 24"	12° 28' 12"	2500	980
Kulmesk	38° 58' 48"	11° 45' 36"	2400	650
Lalibela	39° 1' 48"	12° 1' 12"	2500	790
Sekota	39° 1' 12"	12° 23' 24	2150	635
Nefas Mewcha	38° 27' 0"	11° 42' 0"	3150	1112
Axum	38° 44' 24"	14° 7' 12"	2150	705
Gonder	37° 38' 24"	12° 39' 36"	2130	1100
Yechila	38° 51' 36"	13° 10' 48"	1575	558
Mytsebri	38° 15' 0"	13° 33' 36	1430	1280
Hawzen	39° 25' 48"	13° 58' 4	2250	519
Aguylae	39° 35' 24"	13° 41' 24"	1980	480
Atsbi	39° 44' 24"	13° 52' 12"	2670	600
Mugulat	39° 25' 12"	14° 15' 0"	3200	720
Endabaguna	38° 12' 0"	13° 57' 0"	1760	900
Samre	39° 12' 0"	13° 11' 2	1960	570

Table A-2: Comparison between ground and satellite daily rainfall - PBias (%)

Station name	ARC2	CHIRPS	CMap	CMorph	GPCP	PERSIANN	RFEv2	TRMM
Adi-Arkay	-93	-4	-73	53	-91	-88	-36	80
Ambagiorgis	-10	-35	-97	64	-61	25	-17	73
Debark	-87	-24	-76	37	-85	-55	-44	70
Debre Tabor	-95	-22	-96	11	-93	-47	-31	55
Abi-Adi	-59	25	-100	19	-84	-89	-28	25
Adigrat	-30	-25	-90	58	-3	24	-7	38
Adigudom	-20	-17	185	21	-1	38	3	38
Adishehu	-51	16	33	62	200	-50	36	-23
Adwa	-13	-1	-23	42	37	-52	-34	14
Dengolet	-24	25	32	3	-37	-4	-30	-3
Edaga Hamus	-44	-17	-91	10	-52	-32	40	23
Hagere Selam	-66	-21	91	5	-28	-20	-88	-6
Maichew	-61	13	-37	7	-88	46	46	27
Mykinetal	-20	26	16	23	80	-4	-54	20
Mekele AP	-27	-63	-64	-5	21	10	-16	15
Senkata	-23	-11	15	-6	13	-4	-33	6
Shire	-18	7	104	-7	-89	-17	-22	13
Wukuro	-5	-15	-83	57	-8	15	-12	-27
Korem	-12	-32	-58	231	-67	-55	30	34
Gonder	-6	-12	-4	-1	-62	-27	-71	-24
Yechila	-8	-17	-46	6	14	-47	-10	-8
Mytsebri	-5	-1	-93	-13	-88	-42	-1	-17
Hawzen	-2	7	-98	34	17	-18	-1	23
Aguylae	-5	-40	-92	62	-51	15	-7	-6
Atsbi	-15	-38	-97	-15	-25	-64	-13	-39
Mugulat	-33	-30	-76	-7	-97	-66	-5	42
Endabaguna	-29	-19	22	-5	-38	-38	-16	2
Samre	-2	5	55	3	-34	-22	-22	8
Kulmesk	-94	-14	23	-4	2	-20	1	38
Lalibela	-11	-13	23	-4	-27	-84	0	-24
Sekota	-11	-10	84	20	52	45	-2	-1
Nefas Mewcha	-86	-32	-22	-5	-89	-26	-6	-40
Dabat	-23	-27	-84	-59	-5	80	-57	86
Axum	-28	-16	-12	35	-33	-43	-28	23
Mean	-33	-13	-24	22	-27	-21	-16	17
SD	30	20	72	46	60	41	29	37

Table A-3: Comparison between ground and satellite daily rainfall - correlation coefficient (*r*)

Station name	ARC2	CHIRPS	CMap	CMorph	GPCP	PERSIANN	RFEv2	TRMM
Adi-Arkay	0.20	0.43	0.13	0.20	0.07	0.34	0.47	0.32
Ambagiorgis	0.09	0.36	0.12	0.09	0.10	0.21	0.34	0.25
Debark	0.17	0.36	0.05	0.17	0.10	0.23	0.43	0.28
D/tabour	0.19	0.39	0.03	0.09	0.10	0.24	0.22	0.57
Abi-Adi	0.26	0.58	0.00	0.36	0.03	0.42	0.53	0.41
Adigrat	0.34	0.48	0.22	0.34	0.03	0.42	0.50	0.47
Adigudom	0.33	0.53	0.22	0.33	0.01	0.33	0.49	0.34
Adishehu	0.28	0.48	0.02	0.28	0.29	0.35	0.43	0.54
Adwa	0.24	0.58	0.09	0.44	0.29	0.44	0.61	0.59
Dengolet	0.30	0.44	0.20	0.30	0.22	0.38	0.67	0.61
Edaga Hamus	0.25	0.53	0.22	0.25	0.30	0.25	0.44	0.56
Hagere Selam	0.24	0.49	0.03	0.24	0.25	0.16	0.43	0.39
Maichew	0.28	0.45	0.19	0.28	0.10	0.28	0.37	0.42
Maykental	0.31	0.59	0.21	0.41	0.28	0.49	0.57	0.64
Mekele AP	0.32	0.59	0.21	0.32	0.22	0.36	0.62	0.49
Senkata	0.31	0.61	0.21	0.31	0.28	0.49	0.59	0.48
Shire	0.41	0.58	0.22	0.41	0.23	0.42	0.62	0.60
Wukuro	0.33	0.49	0.22	0.33	0.37	0.40	0.68	0.58
Korem	0.12	0.43	0.25	0.12	0.15	0.37	0.45	0.55
Gonder	0.21	0.45	0.11	0.21	0.14	0.29	0.62	0.37
Yechila	0.36	0.59	0.22	0.36	0.25	0.40	0.57	0.59
Mytsebri	0.24	0.60	0.02	0.34	0.22	0.30	0.52	0.52
Hawzen	0.24	0.60	0.16	0.24	0.30	0.41	0.45	0.58
Aguylae	0.19	0.39	0.13	0.19	0.19	0.30	0.47	0.57
Atsbi	0.19	0.47	0.16	0.09	0.20	0.27	0.43	0.40
Mugulat	0.17	0.39	0.05	0.11	0.19	0.25	0.49	0.40
Endabaguna	0.26	0.52	0.04	0.26	0.30	0.27	0.65	0.62
Samre	0.11	0.58	0.24	0.41	0.25	0.27	0.57	0.63
Kulmesk	0.21	0.48	0.11	0.21	0.23	0.23	0.56	0.47
Lalibela	0.31	0.61	0.21	0.31	0.26	0.26	0.61	0.58
Sekota	0.38	0.44	0.23	0.38	0.16	0.31	0.52	0.47
Nefas Mewcha	0.16	0.39	0.01	0.16	0.09	0.23	0.51	0.49
Dabat	0.06	0.38	0.17	0.06	0.04	0.24	0.39	0.25
Axum	0.19	0.56	0.13	0.33	0.22	0.40	0.58	0.49
Mean	**0.24**	**0.50**	**0.14**	**0.26**	**0.19**	**0.32**	**0.51**	**0.49**
SD	**0.08**	**0.08**	**0.08**	**0.11**	**0.10**	**0.08**	**0.10**	**0.11**

Table A-4: Comparison between ground and satellite daily rainfall - RMSE (mm/day)

Station name	ARC2	CHIRPS	CMap	CMorph	GPCP	PERSIANN	RFEv2	TRMM
Adi-Arkay	74	9	25	15	38	20	15	25
Ambagiorgis	23	9	16	10	30	17	11	10
Debark	24	11	15	9	21	21	14	16
Debre Tabor	60	7	15	12	20	15	9	8
Abi-Adi	19	5	23	14	51	8	12	14
Adigrat	8	3	26	6	13	11	4	8
Adigudom	12	3	28	5	14	18	4	13
Adishehu	10	2	18	5	25	25	5	14
Adwa	8	2	12	3	25	23	5	15
Dengolet	12	5	6	4	8	9	6	3
Edaga Hamus	32	4	8	5	8	12	6	8
Hagere Selam	26	5	8	5	6	14	11	11
Maichew	29	5	15	3	4	16	12	12
Maykental	16	7	4	2	11	13	7	5
Mekele AP	10	6	16	10	5	7	14	14
Senkata	12	3	7	4	6	17	5	3
Shire	9	5	7	4	20	11	7	4
Wukuro	14	4	15	9	6	20	8	11
Korem	9	6	5	10	16	18	12	13
Gonder	31	5	8	10	17	6	15	13
Yechila	14	2	4	3	6	12	11	11
Mytsebri	11	5	31	18	32	16	12	9
Hawzen	16	2	8	5	6	11	5	7
Aguylae	22	6	19	12	26	26	8	11
Atsbi	29	8	12	80	8	14	6	14
Mugulat	27	8	15	9	18	25	7	8
Endabaguna	10	4	6	4	15	15	12	10
Samre	7	7	11	4	15	23	8	6
Kulmesk	27	4	6	11	8	23	15	8
Lalibela	9	5	6	4	6	6	8	10
Sekota	27	4	10	4	17	26	10	17
Nefas Mewcha	16	5	7	11	19	21	17	21
Dabat	29	9	17	10	11	18	14	13
Axum	10	7	7	4	7	5	6	14
Mean	20	5	13	10	16	16	9	11
SD	14	2	7	13	11	6	4	5

Table A-5: Comparison between ground and satellite daily rainfall - MAE (mm/day)

Station name	ARC2	CHIRPS	CMap	CMorph	GPCP	PERSIANN	RFEv2	TRMM
Adi-Arkay	47	12	39	14	27	14	10	16
Ambagiorgis	19	11	23	8	11	12	7	7
Debark	34	0	23	10	13	14	12	7
Debre Tabor	25	13	21	9	12	19	8	5
Abi-Adi	6	5	26	6	3	6	11	6
Adigrat	17	7	39	14	10	8	2	6
Adigudom	8	3	15	3	10	16	3	12
Adishehu	5	9	11	3	23	17	3	10
Adwa	12	3	9	6	23	16	5	11
Dengolet	10	15	8	4	5	7	3	3
Edaga Hamus	8	9	13	10	6	7	5	5
Hagere Selam	28	5	15	6	5	14	8	9
Maichew	23	8	8	5	3	15	7	10
Maykental	16	4	6	7	9	11	5	3
Mekele AP	9	7	26	11	3	3	9	8
Senkata	13	23	12	7	5	12	5	2
Shire	14	11	11	5	17	12	4	1
Wukuro	5	6	23	3	6	17	6	10
Korem	18	16	7	8	6	16	7	12
Gonder	11	9	13	6	11	6	9	9
Yechila	9	3	7	3	6	9	7	10
Mytsebri	19	13	48	21	40	14	9	7
Hawzen	7	2	13	5	7	4	4	6
Aguylae	16	6	30	11	8	17	8	9
Atsbi	25	11	19	7	14	7	4	10
Mugulat	18	13	22	8	5	21	6	6
Endabaguna	8	2	9	3	9	11	11	8
Samre	13	7	15	6	8	9	5	4
Kulmesk	1	12	9	9	13	15	13	6
Lalibela	6	4	9	3	13	6	4	7
Sekota	5	1	16	5	9	17	6	13
Nefas Mewcha	19	13	10	9	14	18	11	18
Dabat	29	17	27	10	26	13	10	11
Axum	10	8	11	2	15	4	5	10
Mean	15	8	17	7	12	12	7	8
SD	10	5	10	4	8	5	3	4

Table A-6: Comparison between ground and satellite wet season rainfall - using RMSE (mm/season)

Station name	ARC2	CHIRPS	CMap	CMorph	GPCP	PERSIANN	RFEv2	TRMM
Adi-Arkay	418	79	342	634	514	122	114	321
Ambagiorgis	226	63	122	299	247	87	63	125
Debark	248	71	117	288	294	86	192	82
DebrTabor	271	73	112	276	270	135	103	103
Abi-Adi	250	16	173	474	263	67	130	175
Adigrat	2154	31	197	512	218	21	59	57
Adigudom	155	17	63	179	229	46	55	93
Adishehu	143	85	60	148	507	34	67	155
Adwa	196	26	88	246	5207	22	53	61
Dengolet	162	31	47	104	120	16	45	85
EdagHamus	166	25	58	168	136	39	57	49
Hag/Selam	139	29	63	163	103	51	89	63
Maichew	184	34	36	126	71	61	87	62
Maykental	151	13	31	80	204	25	57	66
Mekele AP	153	37	122	308	75	63	32	55
Senkata	103	18	56	150	107	30	56	76
Shire	84	8	49	135	383	30	63	87
Wukuro	71	37	110	300	87	46	52	59
Korem	141	35	38	95	300	87	87	67
Gonder	161	29	59	163	283	79	109	63
Yechila	61	7	34	89	62	45	26	77
Mytsebri	147	75	233	418	248	85	104	112
Hawzen	154	10	61	164	117	26	48	49
Aguylae	105	37	147	393	173	46	63	52
Atsbi	98	70	89	244	83	69	42	68
Mugulat	123	71	117	288	109	63	50	98
Endabaguna	143	11	47	121	197	76	90	64
Samre	95	31	80	201	184	23	53	73
Kulmesk	141	64	44	251	207	101	43	90
Lalibela	71	22	44	115	121	56	50	124
Sekota	118	6	75	204	198	65	68	87
Nefasmewch	92	62	51	126	307	113	125	177
Dabat	146	101	130	356	137	79	113	165
Axum	151	43	55	146	120	36	58	56
Mean	212	40	93	234	349	60	74	94
SD	350	26	66	132	866	31	34	54

Table A-7: Comparison between ground and satellite wet season rainfall - MAE (mm/season)

Station name	ARC 2	CHIRP S	CMap	CMorph	GPCP	PERSIAN N	RFEv2	TRMM
Adi-Arkay	318	51	503	517	473	77	93	240
Ambagiorgis	130	41	152	149	104	65	32	96
Debark	134	46	250	144	109	76	88	40
Debre Tabor	238	48	256	138	107	104	60	93
Abi-Adi	105	10	374	383	218	40	29	110
Adigrat	45	20	435	456	198	15	24	33
Adigudom	67	11	127	150	222	16	25	67
Adishehu	55	55	78	104	432	31	27	57
Adwa	44	10	218	163	132	11	29	64
Dengolet	69	20	100	52	87	12	25	48
Edaghamus	78	16	125	84	98	30	29	27
H/Selam	145	19	127	81	81	26	67	52
Maichew	63	22	107	52	23	28	75	58
Maykental	32	8	43	40	157	20	34	68
Mekele AP	57	84	252	154	60	56	25	45
Senkata	66	22	124	75	99	23	35	50
Shire	53	15	121	67	309	12	45	53
Wukuro	60	24	247	250	72	39	47	56
Korem	53	23	56	47	287	33	75	55
Gonder	173	19	55	82	191	34	93	53
Yechila	50	5	43	44	32	30	22	55
Mytsebri	61	49	423	309	166	74	99	86
Hawzen	71	57	126	82	93	20	29	37
Aguylae	122	124	263	297	135	39	38	50
Atsbi	165	45	235	122	40	41	37	57
Mugulat	152	96	250	144	42	55	46	83
Endabaguna	56	77	120	60	178	60	79	47
Samre	37	60	124	101	177	16	51	72
Kulmesk	149	42	59	115	143	84	33	87
Lalibela	51	34	89	97	96	32	49	92
Sekota	52	67	187	182	191	40	65	74
Nefas Mewcha	88	41	106	113	101	98	108	105
Dabat	163	66	256	178	98	71	91	140
Axum	56	28	124	135	89	22	37	59
Mean	96	40	181	152	148	42	51	71
SD	63	28	118	116	103	26	26	38

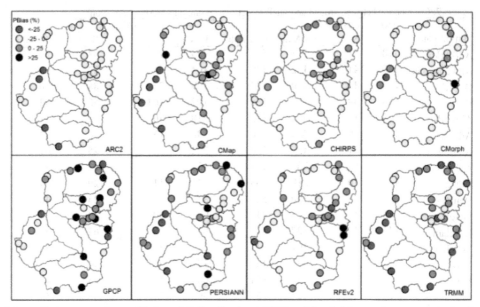

Figure A-1: Comparison of the spatial distribution of PBias (%) of eight products over the basin during the wet season

APPENDIX B (CHAPTER 4)

Table B-1. Summary results of MK, Z statistics on monthly rainfall trends. Negative/positive Z value indicates a decreasing/increasing trend and in bold a statistical significant trend at 5 % confidence level (Z= ±1.96).

Station	Statistics	Jan	Feb	Mar	Apr	May	Jun	Jul	Aug	Sep	Oct	Nov	Dec
Makele	Z	-1.04	-0.57	0.54	-1.56	0.44	1.07	-0.38	-1.45	-1.40	**2.10**	1.67	**5.45**
(AP)	P	0.30	0.56	0.60	0.13	0.66	0.30	0.57	0.39	0.16	**0.04**	0.1	**0.00**
Adigrat	Z	0.95	-0.66	0.32	-0.23	1.16	0.48	-1.18	-1.07	-1.41	0.56	-0.17	**1.99**
	P	0.34	0.51	0.79	0.82	0.25	0.65	0.24	0.33	0.39	0.58	-0.09	**0.04**
Mychew	Z	1.90	**-2.86**	-1.14	-0.56	1.13	0.94	0.32	0.35	0.26	0.68	2.24	-0.28
	P	0.06	**0.02**	0.26	0.58	0.26	0.35	0.75	0.73	0.80	0.50	**0.03**	0.78
Shire	Z	0.82	-0.64	1.32	-0.38	1.32	**2.64**	-0.65	0.28	**2.29**	0.28	0.52	0.18
	P	0.41	0.53	0.19	0.68	0.21	**0.02**	0.63	0.74	**0.02**	0.77	0.61	0.86
D/tabour	Z	**2.44**	**2.11**	1.83	0.72	0.69	-0.04	-1.90	-1.93	-0.55	0.14	**2.17**	**3.59**
	P	**0.05**	**0.04**	0.06	0.47	0.48	0.95	0.07	0.07	0.54	0.21	**0.03**	**0.00**
Mykinetal	Z	**-2.95**	**-2.40**	-0.96	-1.90	-0.50	0.49	-1.79	1.21	**2.60**	-0.72	-1.19	**-2.36**
	P	**0.003**	**0.02**	0.34	0.06	0.62	0.55	0.08	0.20	**0.01**	0.50	0.24	**0.04**
Gonder	Z	-1.49	0.15	-1.41	0.59	0.25	1.01	0.28	0.13	0.52	0.17	-1.66	-1.79
	P	0.14	0.88	0.16	0.54	0.70	0.41	0.78	0.87	0.66	0.87	0.10	0.08
Adigudem	Z	**4.98**	**-2.56**	0.22	0.98	-0.47	-0.68	0.36	0.21	0.24	-0.23	**3.95**	**5.26**
	P	**0.001**	**0.01**	0.83	0.09	0.65	0.16	0.62	0.16	0.81	0.82	**0.01**	**0.01**
H/selam	Z	0.00	-1.13	-1.65	**-2.99**	0.48	-0.27	-0.62	0.19	0.12	-0.51	-0.58	1.42
	P	0.98	**0.01**	0.23	0.68	0.65	0.75	0.11	0.80	0.92	0.56	0.16	**0.03**
Hawzen	Z	-0.48	-1.60	-0.51	-0.55	-0.42	0.08	-0.08	0.37	0.07	-0.91	0.00	-1.13
	P	0.11	0.06	0.55	0.58	0.40	0.93	0.92	0.40	0.92	0.36	1.0	0.26
Wukro	Z	**-6.15**	**-3.79**	-1.48	-0.90	-1.95	1.11	0.10	1.10	-1.09	-0.88	-1.30	**-2.60**
	P	**0.001**	**0.01**	0.14	0.35	0.06	0.22	0.75	0.22	0.38	0.38	0.20	**0.01**
Abiadi	Z	-1.56	-1.90	**-2.27**	**-2.27**	-1.85	1.78	0.72	0.32	1.73	-0.58	-0.40	**2.98**
	P	0.12	0.06	**0.02**	**0.02**	0.06	0.06	0.10	0.14	0.06	0.56	0.71	**0.00**
Demgolat	Z	-1.47	**-3.44**	0.05	-0.08	1.09	0.03	0.13	1.82	0.01	**-2.12**	-0.48	**-2.42**
	P	0.14	**0.01**	0.95	0.86	0.31	0.97	0.83	0.06	0.99	**0.03**	0.63	**0.02**
E/hamus	Z	-1.63	**-1.10**	**-2.45**	-1.62	-1.02	-0.95	0.21	-0.64	-0.59	-0.86	0.33	-1.58
	P	0.10	**0.04**	**0.02**	0.08	0.31	0.39	0.85	0.53	0.56	0.39	0.74	0.08
Adwa	Z	0.25	-0.76	0.24	0.33	1.01	0.96	0.81	**2.19**	1.32	0.35	1.13	0.48
	P	0.68	0.45	0.81	0.74	0.30	0.29	0.23	**0.01**	0.12	0.73	0.27	0.09
Axum	Z	-1.64	-3.52	-1.09	0.20	0.12	0.61	-0.72	0.91	2.25	-1.26	-0.02	0.53
	P	0.10	**0.00**	0.32	0.81	0.83	0.43	0.48	0.26	**0.03**	0.21	0.98	0.60
Debark	Z	1.34	**-2.94**	1.70	0.31	0.82	1.20	1.92	0.51	-0.43	-0.97	-1.11	0.17
	P	0.18	**-0.01**	0.12	0.61	0.44	0.09	0.06	0.36	0.51	0.33	0.29	0.87
Lalibela	Z	1.10	1.74	0.85	-0.22	1.28	1.32	1.17	2.59	4.41	0.71	0.06	0.67
	P	0.27	**0.05**	0.40	0.83	0.23	0.20	0.11	**0.01**	**0.01**	0.48	0.95	**0.01**
Samre	Z	-6.68	-1.22	-0.54	-0.14	-0.68	-2.21	-0.08	-0.04	0.89	-0.29	-1.04	-2.17
	P	**0.001**	**0.03**	0.59	0.90	0.50	0.03	0.91	0.95	0.37	0.77	0.30	**0.03**

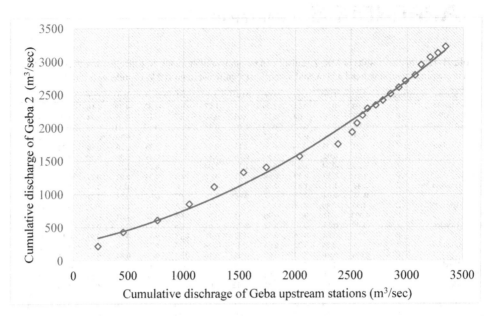

Figure B-1. Comparison of cumulative discharge of Geba 2 with the cumulative discharge of three (Siluh, Genfel and Agula) upstream stations

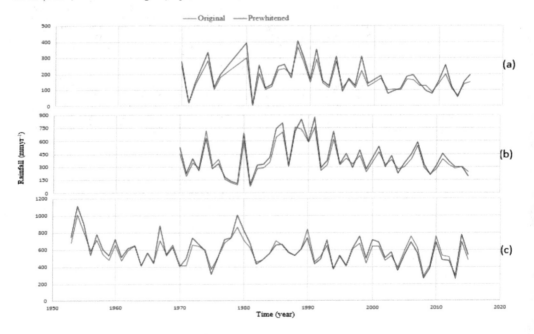

Figure B-2: Comparison of rainfall data before and after removal of auto-correlation: (a) monthly rainfall in Adigrat for August, (b) main rainy season in Adigrat, (c) annual rainfall in Mekelle (AP)

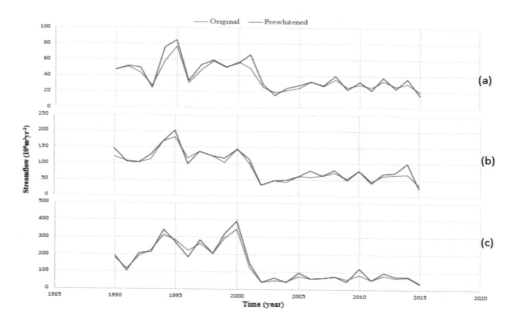

Figure B-3: Comparison of streamflow data before and after removal of auto-correlation at Geba 1 station: (a) monthly for August, (b) main rainy season, (c) annual

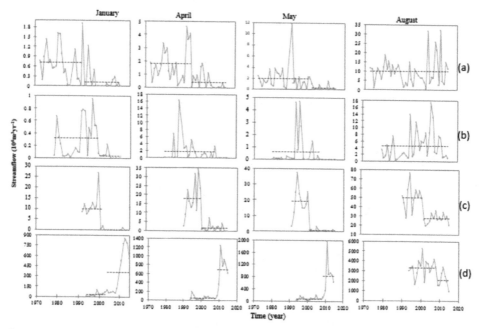

Figure B-4. Pettitt homogeneity test on monthly streamflow for selected stations (a) Siluh, (b) Illala, (c) Geba 1, (d) Emabamadre.

APPENDIX C (CHAPTER 5)

Table C-1: Confusion matrix for 2001, 1989, and 1972 classification maps

Classification result (2001)	Reference (field) data										User's acc. (%)
	WATE	AGRI	BARE	FORE	PLAN	WOOD	BUSH	GRAS	URBA	row totals	
WATE	43	0	0	0	0	0	0	0	0	43	100.0
AGRI	0	473	32	0	1	0	17	17	2	542	87.3
BARE	0	11	234	1	2	0	6	18	0	272	86.0
FORE	0	2	0	163	15	3	0	4	0	187	87.2
PLAN	0	1	0	12	186	5	2	2	0	208	89.4
WOOD	0	0	1	4	1	181	8	0	0	195	92.8
BUSH	0	17	1	0	2	28	274	3	0	325	84.3
GRAS	0	7	1	3	0	0	13	235	0	259	90.7
URBAN	0	1	0	0	1	0	1	0	106	109	97.2
Column totals	43	512	269	183	208	217	321	279	108	2140	
Producer's Accuracy (%)	100	92.4	87.0	89.1	89.4	83.4	85.4	84.2	98.1		

Over all accuracy = 88.5%, Kappa coefficient = 85.3%

1989											
WATE	20	0	0	0	0	0	0	0	0	20	100.0
AGRI	0	453	30	1	8	8	24	23	2	549	82.5
BARE	2	27	285	3	8	9	8	7	0	349	81.7
FORE	0	0	0	179	16	14	0	2	0	211	84.8
PLAN	0	0	0	29	149	1	0	0	0	179	83.2
WOOD	0	1	1	10	0	157	6	1	0	176	89.2
BUSH	0	30	3	2	6	6	302	5	1	355	85.1
GRAS	0	8	2	8	2	1	0	243	0	264	92.0
URBAN	0	2	0	0	0	0	0	0	35	37	94.6
Column totals	22	521	321	232	189	196	340	281	38	2140	
Producer's Accuracy (%)	90.9	86.9	88.8	77.2	78.8	80.1	88.8	86.5	92.1		

Over all accuracy = 88.5%, Kappa coefficient = 85.3%

1972											
WATE	43	0	0	0	0	0	0	0	0	43	100.0
AGRI	0	473	32	0	1	0	17	17	2	542	87.3
BARE	0	11	234	1	2	0	6	18	0	272	86.0
FORE	0	2	0	163	15	3	0	4	0	187	87.2
PLAN	0	1	0	12	186	5	2	2	0	208	89.4
WOOD	0	0	1	4	1	181	8	0	0	195	92.8
BUSH	0	17	1	0	2	28	274	3	0	325	84.3
GRAS	0	7	1	3	0	0	13	235	0	259	90.7
URBAN	0	1	0	0	1	0	1	0	106	109	97.2
Column totals	43	512	269	183	208	217	321	279	108	2140	
Producer's Accuracy (%)	100	92.4	87.0	89.1	89.4	83.4	85.4	84.2	98.1		

Over all accuracy = 88.5%, Kappa coefficient = 85.3%

Table C-2: *LULC transition matrix for 1972-1989 and 1972-2014, area (km²)*

		To final state in 2014											
		AGRI	BUSH	BARE	GRAS	WOOD	FORE	URBE	PLAN	WATE	totals		
	AGRI	1486.5	326.4	103.0	151.4	3.9	0.1	11.1	27.1	1.5	2110.8	624.3	1017.8	0.4
	BUSH	237.0	757.6	75.9	96.7	120.6	20.8	2.3	33.4	1.2	1345.5	587.9	1175.8	0.8
	BARE	136.9	187.6	309.8	78.7	64.5	3.0	0.0	18.5	1.7	800.7	490.9	501.4	1.6
Initial state from 2001	GRAS	75.7	107.5	31.0	90.0	16.1	13.0	0.4	13.5	0.3	347.5	257.5	515.0	2.9
	WOOD	42.6	55.1	30.7	15.7	62.6	22.0	0.0	31.4	0.1	260.1	197.5	394.9	3.2
	FORE	4.4	1.9	3.4	7.7	21.3	12.7	0.0	1.3	1.3	53.9	41.2	82.3	3.2
	URBE	0.0	0.0	0.0	0.0	0.0	0.0	30.5	0.0	0.0	30.5	0.0	0.0	0.0
	PLAN	11.7	5.6	6.6	7.5	17.6	6.6	3.7	68.9	1.6	129.7	60.8	121.6	
	WATE	0.6	0.0	0.2	0.3	0.0	0.0	0.0	0.3	4.9	6.2	1.4	2.7	
	totals	1995.4	1441.6	560.5	447.8	306.7	78.2	47.9	194.3	12.5	5084.8	2261.4	1905.8	0.8
	gain	508.9	684.0	250.7	357.9	244.1	65.5	17.4	125.4	7.6	2261.4			
	Ng	115.4	96.1	240.2	100.4	46.6	24.3	17.4	64.6	6.3	355.6			
	Tg	1133.2	1271.9	741.6	615.4	441.5	106.6	17.4	186.2	9.0	2261.4			
	Gp	0.3	0.9	0.8	4.0	3.9	5.2	0.6	1.8	1.6	0.8			
	Gp	0.6	0.8	0.8	6.4	4.2	1.7	1.5	3.7	NA	0.9			

		To final state in 2001											
		AGRI	BUSH	BARE	GRAS	WOOD	FORE	URBE	PLAN	WATE	totals		
	AGRI	1339.201	166.85	188.78	88.42	0.25	0	14.58	24.31	0.52	1822.9	483.7	967.4	0.4
	BUSH	142.23	764.22	96.45	107.46	161.6	2.00	0.52	18.47	0.29	1293.2	529.0	1058.0	0.7
	BARE	470.14	245.62	442.23	65.75	0.25	0.9	0	22.27	3.63	1250.8	808.6	716.5	1.8
Initial state from 1989	GRAS	90.99	34.39	39.82	46.85	1.94	1.42	0	21.14	0.08	236.6	189.8	379.6	4.1
	WOOD	52.88	83.6	15.9	19.23	49.84	28.53	0	14.77	0.16	264.9	215.1	420.6	4.3
	FORE	11.21	46.34	10.13	14.37	42.24	19.95	0	1.76	0.4	146.4	126.5	67.4	6.3
	URBE	0	0	0	0	0	0	12.16	0	0	12.2	0.0	0.0	0.0
	PLAN	4.05	4.44	7.04	5.4	4	0.86	3.2	27.43	1.25	57.7	30.2	60.5	1.1
	WATE	0.02	0	0.13	0	0	0	0	0	0	0.2	0.2	0.3	
	totals	2110.721	1345.46	800.48	347.48	260.12	53.66	30.46	130.15	6.33	5084.9	2383.0	1835.1	0.9
	gain	771.5	581.2	358.3	300.6	210.3	33.7	18.3	102.7	6.3	2383.0			
	Ng	287.8	52.2	450.3	110.9	4.8	92.7	18.3	72.5	6.2	547.8			
	Tg	1255.2	1110.3	1166.8	490.4	425.4	160.2	18.3	133.0	6.5	2383.0			
	Gp	0.6	0.8	0.8	6.4	4.2	1.7	1.5	3.7	NA	0.9			

APPENDIX D (CHAPTER 6)

Table D-1: wflow_sbm Calibrated parameter value At Geba 1 catchment. The columns represent the land use type (1=Agriculture, 2=Bareland, 3= bushes and shrubs, 4=Forest, 5= Grassland, 6=water bod, 7=Wood land)), sub-catchment, and soil (1=Leptosol,2=Cambisol, 3= Arnosol ,4= arnosol, 5=Vertisol, 6=Luvisol, 7=Calcisol) respectively. The last column is the value assigned based on the interaction of the first three columns.

CanopyGapFraction				EvorR			
1	[0,>	[0,>	0.5	1	[0,>	[0,>	0.1
2	[0,>	[0,>	0.9	2	[0,>	[0,>	0
3	[0,>	[0,>	0.5	3	[0,>	[0,>	0.1
4	[0,>	[0,>	0.3	4	[0,>	[0,>	0.2
6	[0,>	[0,>	0.9	6	[0,>	[0,>	0
7	[0,>	[0,>	0.4	7	[0,>	[0,>	0.1

RootingDepth				N			
1	[0,>	[0,>	200	1	[0,>	[0,>	0.7
2	[0,>	[0,>	300	2	[0,>	[0,>	0.42
3	[0,>	[0,>	800	3	[0,>	[0,>	0.8
4	[0,>	[0,>	1500	4	[0,>	[0,>	0.65
5	[0,>	[0,>	200	5	[0,>	[0,>	0.7
6	[0,>	[0,>	0	6	[0,>	[0,>	0.12
7	[0,>	[0,>	1300	7	[0,>	[0,>	0.8

MaxCanopyStorage				thetaR			
1	[0,>	[0,>	0.3	[0,>	[0,>	1	0.08
2	[0,>	[0,>	0	[0,>	[0,>	2	0.1
3	[0,>	[0,>	0.19	[0,>	[0,>	3	0.08
4	[0,>	[0,>	0.36	[0,>	[0,>	4	0.07
5	[0,>	[0,>	0.23	[0,>	[0,>	5	0.13
6	[0,>	[0,>	0	[0,>	[0,>	6	0.09
7	[0,>	[0,>	0.39	[0,>	[0,>	7	0.08

InfiltCapPath.tbl				SoilMinThickness			
[0,>	[0,>	1	28	[0,>	[0,>	1	10
[0,>	[0,>	2	30	[0,>	[0,>	2	100
[0,>	[0,>	3	5	[0,>	[0,>	3	300
[0,>	[0,>	4	5	[0,>	[0,>	4	250
[0,>	[0,>	5	5	[0,>	[0,>	5	200
[0,>	[0,>	6	20	[0,>	[0,>	6	50
[0,>	[0,>	7	5	[0,>	[0,>	7	40

SoilThikness				InfiltCapSoil			
[0,>	[0,>	1	250	[0,>	[0,>	1	127
[0,>	[0,>	2	2000	[0,>	[0,>	2	160
[0,>	[0,>	3	1500	[0,>	[0,>	3	200
[0,>	[0,>	4	1200	[0,>	[0,>	4	95
[0,>	[0,>	5	1800	[0,>	[0,>	5	27
[0,>	[0,>	6	1500	[0,>	[0,>	6	30
[0,>	[0,>	7	1000	[0,>	[0,>	7	130

Ksat				thetaS			
[0,>	[0,>	1	500	[0,>	[0,>	1	0.3
[0,>	[0,>	2	708	[0,>	[0,>	2	0.4
[0,>	[0,>	3	1500	[0,>	[0,>	3	0.2
[0,>	[0,>	4	400	[0,>	[0,>	4	0.2
[0,>	[0,>	5	450	[0,>	[0,>	5	0.5
[0,>	[0,>	6	1200	[0,>	[0,>	6	0.2
[0,>	[0,>	7	900	[0,>	[0,>	7	0.2

M				PathFrac			
[0,>	[0,>	1	60	[0,>	[0,>	1	0.12
[0,>	[0,>	2	640	[0,>	[0,>	2	0.03
[0,>	[0,>	3	560	[0,>	[0,>	3	0.08
[0,>	[0,>	4	450	[0,>	[0,>	4	0.1
[0,>	[0,>	5	450	[0,>	[0,>	5	0.06
[0,>	[0,>	6	170	[0,>	[0,>	6	0.04
[0,>	[0,>	7	150	[0,>	[0,>	7	0.06

Beta				CapScale			
[0,>	[0,>	[0,>	0.6	[0,>	[0,>	[0,>	100

N_River				RunoffGeneratingGWPerc			
[0,>	[0,>	[0,>	0.04	[0,>	[0,>	[0,>	0.1

Table D-2: wflow_sbm Calibrated parameter value at Geba 2 Sub_catchment (reverse modelling). The columns represent the land use type, sub-catchment, and soil, respectively. The last column is the value assigned based on the interaction of the first three columns.

CanopyGapFraction				EvorR			
1	[0,>	[0,>	0.5	1	[0,>	[0,>	0.1
2	[0,>	[0,>	0.9	2	[0,>	[0,>	0
3	[0,>	[0,>	0.4	3	[0,>	[0,>	0.2
4	[0,>	[0,>	0.3	4	[0,>	[0,>	0.1
6	[0,>	[0,>	0.9	6	[0,>	[0,>	0
7	[0,>	[0,>	0.3	7	[0,>	[0,>	0.2

RootingDepth				N			
1	[0,>	[0,>	200	1	[0,>	[0,>	0.73
2	[0,>	[0,>	400	2	[0,>	[0,>	0.39
3	[0,>	[0,>	750	3	[0,>	[0,>	0.8
4	[0,>	[0,>	1800	4	[0,>	[0,>	0.63
5	[0,>	[0,>	200	5	[0,>	[0,>	0.6
6	[0,>	[0,>	0	6	[0,>	[0,>	0.10
7	[0,>	[0,>	1600	7	[0,>	[0,>	0.7

MaxCanopyStorage				thetaR			
1	[0,>	[0,>	0.3	[0,>	[0,>	1	0.07
2	[0,>	[0,>	0	[0,>	[0,>	2	0.1
3	[0,>	[0,>	0.23	[0,>	[0,>	3	0.08
4	[0,>	[0,>	0.38	[0,>	[0,>	4	0.06
5	[0,>	[0,>	0	[0,>	[0,>	5	0.12
6	[0,>	[0,>	0.23	[0,>	[0,>	6	0.08
7	[0,>	[0,>	0.41	[0,>	[0,>	7	0.08

InfiltCapPath.tbl				SoilMinThikness			
[0,>	[0,>	1	27	[0,>	[0,>	1	12
[0,>	[0,>	2	31	[0,>	[0,>	2	120
[0,>	[0,>	3	5	[0,>	[0,>	3	300
[0,>	[0,>	4	4	[0,>	[0,>	4	250
[0,>	[0,>	5	5	[0,>	[0,>	5	200
[0,>	[0,>	6	21	[0,>	[0,>	6	60
[0,>	[0,>	7	4	[0,>	[0,>	7	40

SoilThikness				InfiltCapSoil			
[0,>	[0,>	1	250	[0,>	[0,>	1	127
[0,>	[0,>	2	2000	[0,>	[0,>	2	160
[0,>	[0,>	3	1500	[0,>	[0,>	3	200
[0,>	[0,>	4	1200	[0,>	[0,>	4	95
[0,>	[0,>	5	1800	[0,>	[0,>	5	27
[0,>	[0,>	6	1500	[0,>	[0,>	6	30
[0,>	[0,>	7	1000	[0,>	[0,>	7	130

Ksat				thetaS			
[0,>	[0,>	1	600	[0,>	[0,>	1	0.2
[0,>	[0,>	2	728	[0,>	[0,>	2	0.3
[0,>	[0,>	3	1600	[0,>	[0,>	3	0.3
[0,>	[0,>	4	450	[0,>	[0,>	4	0.2
[0,>	[0,>	5	450	[0,>	[0,>	5	0.4
[0,>	[0,>	6	1400	[0,>	[0,>	6	0.3
[0,>	[0,>	7	1000	[0,>	[0,>	7	0.1

M				PathFrac			
[0,>	[0,>	1	100	[0,>	[0,>	1	0.11
[0,>	[0,>	2	840	[0,>	[0,>	2	0.04
[0,>	[0,>	3	620	[0,>	[0,>	3	0.07
[0,>	[0,>	4	540	[0,>	[0,>	4	0.1
[0,>	[0,>	5	630	[0,>	[0,>	5	0.05
[0,>	[0,>	6	160	[0,>	[0,>	6	0.04
[0,>	[0,>	7	140	[0,>	[0,>	7	0.05

Beta				CapScale			
[0,>	[0,>	[0,>	0.6	[0,>	[0,>	[0,>	100

N_River				RunoffGeneratingGWPerc			
[0,>	[0,>	[0,>	0.04	[0,>	[0,>	[0,>	0.1

Table D-3: wflow_sbm Calibrated parameter value at Siluh sub_catchment (Reverse modelling). The columns represent the land use type, sub-catchment, and soil, respectively. The last column is the value assigned based on the interaction of the first three columns.

CanopyGapFraction				EvorR			
1	[0,>	[0,>	0.3	1	[0,>	[0,>	0.1
2	[0,>	[0,>	0.9	2	[0,>	[0,>	0
3	[0,>	[0,>	0.5	3	[0,>	[0,>	0.1
4	[0,>	[0,>	0.3	4	[0,>	[0,>	0.2
6	[0,>	[0,>	0.8	6	[0,>	[0,>	0
7	[0,>	[0,>	0.4	7	[0,>	[0,>	0.1

RootingDepth				N			
1	[0,>	[0,>	200	1	[0,>	[0,>	0.73
2	[0,>	[0,>	250	2	[0,>	[0,>	0.39
3	[0,>	[0,>	700	3	[0,>	[0,>	0.8
4	[0,>	[0,>	1600	4	[0,>	[0,>	0.63
5	[0,>	[0,>	200	5	[0,>	[0,>	0.6
6	[0,>	[0,>	0	6	[0,>	[0,>	0.10
7	[0,>	[0,>	1600	7	[0,>	[0,>	0.7

MaxCanopyStorage				thetaR			
1	[0,>	[0,>	0.3	[0,>	[0,>	1	0.07
2	[0,>	[0,>	0	[0,>	[0,>	2	0.1
3	[0,>	[0,>	0.23	[0,>	[0,>	3	0.08
4	[0,>	[0,>	0.38	[0,>	[0,>	4	0.06
5	[0,>	[0,>	0.23	[0,>	[0,>	5	0.12
6	[0,>	[0,>	0.38	[0,>	[0,>	6	0.08
7	[0,>	[0,>	0	[0,>	[0,>	7	0.08

InfiltCapPath.tbl				SoilMinThikness			
[0,>	[0,>	1	29	[0,>	[0,>	1	8
[0,>	[0,>	2	35	[0,>	[0,>	2	130
[0,>	[0,>	3	6	[0,>	[0,>	3	320
[0,>	[0,>	4	4	[0,>	[0,>	4	250
[0,>	[0,>	5	7	[0,>	[0,>	5	250
[0,>	[0,>	6	28	[0,>	[0,>	6	70
[0,>	[0,>	7	6	[0,>	[0,>	7	50

SoilThikness				InfiltCapSoil			
[0,>	[0,>	1	500	[0,>	[0,>	1	123
[0,>	[0,>	2	2000	[0,>	[0,>	2	170
[0,>	[0,>	3	1700	[0,>	[0,>	3	250
[0,>	[0,>	4	1300	[0,>	[0,>	4	25
[0,>	[0,>	5	1800	[0,>	[0,>	5	29
[0,>	[0,>	6	1400	[0,>	[0,>	6	30
[0,>	[0,>	7	1100	[0,>	[0,>	7	123

Ksat				thetaS			
[0,>	[0,>	1	800	[0,>	[0,>	1	0.2
[0,>	[0,>	2	928	[0,>	[0,>	2	0.4
[0,>	[0,>	3	1800	[0,>	[0,>	3	0.4
[0,>	[0,>	4	650	[0,>	[0,>	4	0.2
[0,>	[0,>	5	550	[0,>	[0,>	5	0.6
[0,>	[0,>	6	1600	[0,>	[0,>	6	0.4
[0,>	[0,>	7	1200	[0,>	[0,>	7	0.2

M				PathFrac			
[0,>	[0,>	1	100	[0,>	[0,>	1	0.10
[0,>	[0,>	2	860	[0,>	[0,>	2	0.03
[0,>	[0,>	3	630	[0,>	[0,>	3	0.05
[0,>	[0,>	4	560	[0,>	[0,>	4	0.1
[0,>	[0,>	5	460	[0,>	[0,>	5	0.04
[0,>	[0,>	6	140	[0,>	[0,>	6	0.04
[0,>	[0,>	7	140	[0,>	[0,>	7	0.04

Beta				CapScale			
[0,>	[0,>	[0,>	0.6	[0,>	[0,>	[0,>	100

N_River				RunoffGeneratingGWPerc			
[0,>	[0,>	[0,>	0.04	[0,>	[0,>	[0,>	0.1

Table D-4: wflow_sbm Calibrated parameter value At Geba2 sub_catchment (Forward modelling). The columns represent the land use type, sub-catchment, and soil, respectively. The last column is the value assigned based on the interaction of the first three columns.

CanopyGapFraction				EvorR			
1	[0,>	[0,>	0.3	1	[0,>	[0,>	0.1
2	[0,>	[0,>	0.9	2	[0,>	[0,>	0
3	[0,>	[0,>	0.5	3	[0,>	[0,>	0.1
4	[0,>	[0,>	0.3	4	[0,>	[0,>	0.2
6	[0,>	[0,>	0.8	6	[0,>	[0,>	0
7	[0,>	[0,>	0.3	7	[0,>	[0,>	0.1
RootingDepth				N			
1	[0,>	[0,>	200	1	[0,>	[0,>	0.77
2	[0,>	[0,>	250	2	[0,>	[0,>	0.45
3	[0,>	[0,>	700	3	[0,>	[0,>	0.8
4	[0,>	[0,>	1600	4	[0,>	[0,>	0.71
5	[0,>	[0,>	200	5	[0,>	[0,>	0.63
6	[0,>	[0,>	0	6	[0,>	[0,>	0.10
7	[0,>	[0,>	1600	7	[0,>	[0,>	0.74
MaxCanopyStorage				thetaR			
1	[0,>	[0,>	0.36	[0,>	[0,>	1	0.08
2	[0,>	[0,>	0.12	[0,>	[0,>	2	0.2
3	[0,>	[0,>	0.28	[0,>	[0,>	3	0.11
4	[0,>	[0,>	0.42	[0,>	[0,>	4	0.09
5	[0,>	[0,>	0	[0,>	[0,>	5	0.16
6	[0,>	[0,>	0.24	[0,>	[0,>	6	0.10
7	[0,>	[0,>	0.41	[0,>	[0,>	7	0.09
InfiltCapPath.tbl				SoilMinThickness			
[0,>	[0,>	1	30	[0,>	[0,>	1	13
[0,>	[0,>	2	32	[0,>	[0,>	2	125
[0,>	[0,>	3	4	[0,>	[0,>	3	300
[0,>	[0,>	4	3	[0,>	[0,>	4	250
[0,>	[0,>	5	5	[0,>	[0,>	5	200
[0,>	[0,>	6	26	[0,>	[0,>	6	62
[0,>	[0,>	7	6	[0,>	[0,>	7	55
SoilThikness				InfiltCapSoil			
[0,>	[0,>	1	500	[0,>	[0,>	1	130
[0,>	[0,>	2	2000	[0,>	[0,>	2	180
[0,>	[0,>	3	1700	[0,>	[0,>	3	300
[0,>	[0,>	4	1300	[0,>	[0,>	4	35
[0,>	[0,>	5	1800	[0,>	[0,>	5	31
[0,>	[0,>	6	1400	[0,>	[0,>	6	35
[0,>	[0,>	7	1100	[0,>	[0,>	7	127
Ksat				thetaS			
[0,>	[0,>	1	750	[0,>	[0,>	1	0.2
[0,>	[0,>	2	800	[0,>	[0,>	2	0.3
[0,>	[0,>	3	1650	[0,>	[0,>	3	0.4
[0,>	[0,>	4	550	[0,>	[0,>	4	0.4
[0,>	[0,>	5	550	[0,>	[0,>	5	0.6
[0,>	[0,>	6	1500	[0,>	[0,>	6	0.4
[0,>	[0,>	7	1000	[0,>	[0,>	7	0.3
M				PathFrac			
[0,>	[0,>	1	80	[0,>	[0,>	1	0.09
[0,>	[0,>	2	660	[0,>	[0,>	2	0.03
[0,>	[0,>	3	530	[0,>	[0,>	3	0.06
[0,>	[0,>	4	460	[0,>	[0,>	4	0.08
[0,>	[0,>	5	460	[0,>	[0,>	5	0.03
[0,>	[0,>	6	150	[0,>	[0,>	6	0.03
[0,>	[0,>	7	120	[0,>	[0,>	7	0.02
Beta				CapScale			
[0,>	[0,>	[0,>	0.6	[0,>	[0,>	[0,>	100
N_River				RunoffGeneratingGWPerc			
[0,>	[0,>	[0,>	0.04	[0,>	[0,>	[0,>	0.1

Table D-5: wflow_sbm Calibrated parameter value at Siluh sub_catchment. The columns represent the land use type, sub-catchment, and soil, respectively. The last column is the value assigned based on the interaction of the first three columns.

CanopyGapFraction				EvorR			
1	[0,>	[0,>	0.3	1	[0,>	[0,>	0.1
2	[0,>	[0,>	0.8	2	[0,>	[0,>	0
3	[0,>	[0,>	0.4	3	[0,>	[0,>	0.2
4	[0,>	[0,>	0.2	4	[0,>	[0,>	0.2
6	[0,>	[0,>	0.8	6	[0,>	[0,>	0
7	[0,>	[0,>	0.3	7	[0,>	[0,>	0.2
RootingDepth				**N**			
1	[0,>	[0,>	200	1	[0,>	[0,>	0.76
2	[0,>	[0,>	250	2	[0,>	[0,>	0.40
3	[0,>	[0,>	800	3	[0,>	[0,>	0.82
4	[0,>	[0,>	1600	4	[0,>	[0,>	0.66
5	[0,>	[0,>	200	5	[0,>	[0,>	0.6
6	[0,>	[0,>	0	6	[0,>	[0,>	0.12
7	[0,>	[0,>	1700	7	[0,>	[0,>	0.8
MaxCanopyStorage				**thetaR**			
1	[0,>	[0,>	0.4	[0,>	[0,>	1	0.10
2	[0,>	[0,>	0.1	[0,>	[0,>	2	0.14
3	[0,>	[0,>	0.5	[0,>	[0,>	3	0.10
4	[0,>	[0,>	0.48	[0,>	[0,>	4	0.07
5	[0,>	[0,>	0.26	[0,>	[0,>	5	0.13
6	[0,>	[0,>	0	[0,>	[0,>	6	0.08
7	[0,>	[0,>	0.45	[0,>	[0,>	7	0.10
InfiltCapPath.tbl				**SoilMinThikness**			
[0,>	[0,>	1	30	[0,>	[0,>	1	9
[0,>	[0,>	2	34	[0,>	[0,>	2	130
[0,>	[0,>	3	6	[0,>	[0,>	3	310
[0,>	[0,>	4	4	[0,>	[0,>	4	240
[0,>	[0,>	5	8	[0,>	[0,>	5	260
[0,>	[0,>	6	30	[0,>	[0,>	6	75
[0,>	[0,>	7	6	[0,>	[0,>	7	60
SoilThikness				**InfiltCapSoil**			
[0,>	[0,>	1	500	[0,>	[0,>	1	125
[0,>	[0,>	2	2000	[0,>	[0,>	2	170
[0,>	[0,>	3	1700	[0,>	[0,>	3	250
[0,>	[0,>	4	1300	[0,>	[0,>	4	27
[0,>	[0,>	5	1800	[0,>	[0,>	5	30
[0,>	[0,>	6	1400	[0,>	[0,>	6	28
[0,>	[0,>	7	1100	[0,>	[0,>	7	128
Ksat				**thetaS**			
[0,>	[0,>	1	1000	[0,>	[0,>	1	0.3
[0,>	[0,>	2	1028	[0,>	[0,>	2	0.5
[0,>	[0,>	3	2000	[0,>	[0,>	3	0.5
[0,>	[0,>	4	750	[0,>	[0,>	4	0.3
[0,>	[0,>	5	580	[0,>	[0,>	5	0.6
[0,>	[0,>	6	1800	[0,>	[0,>	6	0.5
[0,>	[0,>	7	1350	[0,>	[0,>	7	0.23
M				**PathFrac**			
[0,>	[0,>	1	150	[0,>	[0,>	1	0.08
[0,>	[0,>	2	1060	[0,>	[0,>	2	0.03
[0,>	[0,>	3	1030	[0,>	[0,>	3	0.04
[0,>	[0,>	4	760	[0,>	[0,>	4	0.2
[0,>	[0,>	5	760	[0,>	[0,>	5	0.03
[0,>	[0,>	6	240	[0,>	[0,>	6	0.02
[0,>	[0,>	7	240	[0,>	[0,>	7	0.02
Beta				**CapScale**			
[0,>	[0,>	[0,>	0.6	[0,>	[0,>	[0,>	100
N_River				**RunoffGeneratingGWPerc**			
[0,>	[0,>	[0,>	0.04	[0,>	[0,>	[0,>	0.1

Table D-6: Mean annual hydrological fluxes (mm/year) of each LULC maps from the reverse modelling approach using climatic data of 2013-2015.

Genfel	LULC_1972	LULC_1989	LULC_2001	LULC_2014
Annual rainfall	520	520	520	520
Storage	62	47	52	69
AET	314	286	303	332
Annual flow	181	216	204	163
Wet season flow	101	157	125	98
Dry season flow	80	59	79	65
Runoff coefficient	0.35	0.42	0.39	0.31
Wet season/annual flow	0.19	0.30	0.24	0.19
Dry season /annual flow	0.44	0.27	0.39	0.40
Ilala				
Annual rainfall	530	530	530	530
Storage	156	125	150	140
AET	302	262	306	329
Annual flow	119	165	95	82
Wet season flow	94	155	80	69
Dry season flow	25	10	15	13
Runoff coefficient	0.22	0.31	0.18	0.15
Wet season/annual flow	0.79	0.94	0.84	0.84
Dry season /annual flow	0.21	0.06	0.16	0.16

Table D-7: Mean annual hydrological fluxes (mm/year) of each LULC maps from the forward modelling approach using climatic data of 1974-1976.

Geba2	LULC_1972	LULC_1989	LULC_2001	LULC_2014
Annual rainfall	770	770	770	770
Storage	178	102	131	154
AET	460	354	458	495
Annual flow	164	336	208	145
Wet season flow	103	283	152	102
Dry season flow	61	54	56	53
Runoff coefficient	0.21	0.44	0.27	0.19
Wet season /annual flow	0.63	0.84	0.73	0.71
Dry season /annual flow	0.37	0.16	0.27	0.36
Siluh				
Annual rainfall	675	675	675	675
Storage	137	90	123	156
AET	408	293	371	388
Annual flow	155	307	204	127
Wet season flow	94	239	153	85
Dry season flow	61	68	52	62
Runoff coefficient	0.23	0.46	0.30	0.19
Wet season /annual flow	0.61	0.78	0.75	0.67
Dry season /annual flow	0.39	0.22	0.25	0.49

Table D-8: Summary of the PLSR models of streamflow (annual, wet & dry season flows), AET and Storage. Result from forward modelling approach in Geba2 sub-catchment for the entire period (1972-2014)

Response Y	R2x	R2y	Q2	Component	% of explained variability in y	Cumulative % of explained variability in y	RMSEcv	Q2cum
Streamflow (Annual, Wet Dry flows)	0.72	0.68	0.82	1	76.20	76.20	12.80	0.6
				2	19.50	95.90	14.80	0.8
AET	0.68	0.74	0.78	1	93.70	93.70	12.40	0.8
Storage	0.95	0.72	0.82	1	94.50	64.50	11.20	83.4

Table D-9: Variable importance of the projection values (VIP) and PLSR for the hydrological components (from forward modelling approach) in Geba2 sub-catchment for the entire study period

Predictors	Streamflow (Annual, wet and dry season flows)			AET		Storage	
	VIP	W*(1)	W*(2)	VIP	W*(1)	VIP	W*(1)
AGRI	1.30	0.27	**-0.48**	0.23	-0.09	1.49	**-0.34**
WOOD	1.09	**-0.33**	**0.41**	0.47	0.18	1.02	**0.39**
FORE	1.17	**-0.44**	0.00	1.08	**0.41**	1.17	**0.44**
BARE	1.22	**0.46**	0.20	1.30	**-0.49**	1.12	**-0.42**
WATE	0.44	-0.16	-0.17	0.80	0.28	0.21	0.08
BUSH	1.18	**-0.45**	-0.27	1.33	**0.50**	1.07	**0.40**
GRAS	1.14	**-0.43**	0.00	1.04	**0.39**	1.16	**0.44**

APPENDIX E (CHAPTER 7)

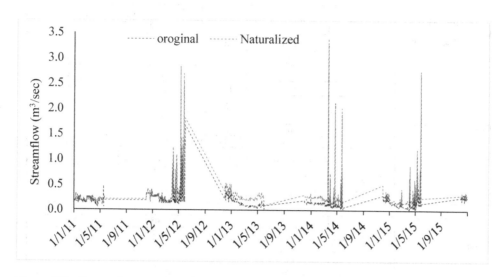

Figure E-1: Comparison of streamflow before and after naturalizing in Agula catchment

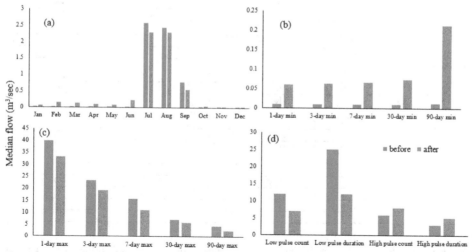

Figure E-2: Comparison of hydrological responses in Genfel catchment before and after catchment management intervention using hydrological alteration parameters: (a) Magnitude of average median monthly flow; (b) Magnitude of median annual average minima flows, (c) Magnitude of median average annual maxima flows and (d) Frequency and duration of average high and low pulses

Table E-1: Model parameter values comparison between the paired catchments.

Model parameter	Land use	Sub-Catchment	Soil	Parameter value	
				Agula catchment	Genfel Catchment
CanopyGapFraction	1	[0,>	[0,>	0.4	0.4
	2	[0,>	[0,>	0.9	0.9
	3	[0,>	[0,>	0.5	0.5
	4	[0,>	[0,>	0.2	0.3
	6	[0,>	[0,>	1.0	1.0
	7	[0,>	[0,>	0.5	0.4
EVOR	1	[0,>	[0,>	0.1	0.1
	2	[0,>	[0,>	0.01	0.01
	3	[0,>	[0,>	0.05	0.05
	4	[0,>	[0,>	0.4	0.3
	6	[0,>	[0,>	0	0
	7	[0,>	[0,>	0.3	0.3
Ksat	[0,>	[0,>	1	800	600
	[0,>	[0,>	2	1208	908
	[0,>	[0,>	3	2300	1700
	[0,>	[0,>	4	780	630
	[0,>	[0,>	5	950	750
	[0,>	[0,>	6	2200	1600
	[0,>	[0,>	7	1400	1100
	[0,>	[0,>	1	70	40
	[0,>	[0,>	2	560	340
	[0,>	[0,>	3	660	400
M	[0,>	[0,>	4	430	250
	[0,>	[0,>	5	430	250
	[0,>	[0,>	6	160	120
	[0,>	[0,>	7	180	140
InfiltCapSoil	[0,>	[0,>	1	98	113
	[0,>	[0,>	2	110	140
	[0,>	[0,>	3	150	210
	[0,>	[0,>	4	65	85
	[0,>	[0,>	5	26	34
	[0,>	[0,>	6	25	30
	[0,>	[0,>	6	80	120
thetaS	[0,>	[0,>	1	0.3	0.1
	[0,>	[0,>	2	0.4	0.2
	[0,>	[0,>	3	0.4	0.2
	[0,>	[0,>	4	0.3	0.1
	[0,>	[0,>	5	0.6	0.3
	[0,>	[0,>	6	0.5	0.3
	[0,>	[0,>	7	0.3	0.2

Note that, the columns represent the land use type, sub-catchment, and soil, respectively. The last column is the value assigned based on the interaction of the first three columns. The land-use representations are: 1.Agriculture, 2. Bareland, 3. Bushes and shrubs, 4. Forest land, 5. Grass land, 6.Water body, 7. Woody vegetation; and the soil representations are: 1. Leptisols, 2. Cambisols, 3. Arnosols, 4. Alisosl, 5.Vertisols, 6. Luvisols and 7.Calsisols. The notation '[0,>' means all values greater than 0.

SAMENVATTING

Het stroomgebied van de Tekeze-Atbara, een van de grootste rivieren van het stroomgebied van de Nijl, wordt gedeeld tussen Ethiopië en Sudan, en is cruciaal voor economische ontwikkeling en ecologische duurzaamheid in de regio. Het stroomopwaartse deel van dit stroomgebied, en met name de bovenloop van het stroomgebied van de Upper Tekeze rivier, is momenteel de focus van de Ethiopische regering op economische ontwikkeling in de droge gebieden van Noord-Ethiopië. De regering heeft zich gecommitteerd aan een ambitieus plan om het voedseltekort van de regio uit te roeien door gebruik te maken van de watervoorraden van het stroombekken voor irrigatie en waterkrachtontwikkeling. Desondanks zijn bodemdegradatie, waterschaarste en inefficiënt gebruik van de beschikbare waterbronnen knelpunten om dit ambitieuze doel te bereiken. Beschikbaarheid van waterbronnen voor economische ontwikkeling in de regio is beïnvloed door verschillende natuurlijke en antropogene factoren. Daarom is het begrijpen van variabiliteit en drijfveren voor verandering van de hydrologie van het stroomgebied van de Upper Tekeze en de implicaties daarvan voor de beschikbaarheid van water van vitaal belang voor beter waterbeheer in de regio.

Het stroomgebied van de Upper Tekeze ($\sim$ 45.000 km^2) wordt niet alleen gekenmerkt door ernstige bodemdegradatie en schaarste aan bodemvocht, maar is ook bekend om zijn recente succesvolle ervaringen met geïntegreerde stroomgebiedsbeheer. Er wordt beweerd dat stroomgebiedsinterventies de waterbeschikbaarheid op verschillende locaties in de bovenste delen van het bekken hebben vergroot. Maar het is niet wetenschappelijk bekend hoe deze door de mens veroorzaakte milieuveranderingen de hydrologische processen beïnvloeden en wat de benedenstroomse gevolgen zijn. Eerdere studies zijn gedaan in experimentele plots of in zeer kleine micro-stroomgebieden, waarvan het moeilijk is om te extrapoleren en implicaties voor het hele stroomgebied te bepalen. Daarom is deze studie gericht op een beter begrip van de impact van antropogene stroomgebiedbeheer dynamieken op de hydrologische processen op verschillende ruimtelijke en temporele schalen en hun benedenstroomse impact. Dit wordt bereikt door gecombineerd gebruik van primaire en secundaire informatie, satellietgegevens, laboratoriumanalyse en beoordeling met behulp van hydrologische modellen.

Satellietproducten voor regenval zijn een belangrijke bron van neerslaginformatie in stroomgebieden waar informatie schaars is, zoals in het Upper Tekeze bekken. Eerst zijn de prestaties van acht veel gebruikte op satelliet gebaseerde schattingen van neerslag (TRMM, CHIRPS, RFEv2, ARC2, PERSIANN, GPCP, CMap en CMorph) geëvalueerd tegen 34 grondmetingen om te bepalen welke producten geschikt zijn voor het Upper Tekeze stroomgebied. De vergelijking van deze producten met de waargenomen regenval werd uitgevoerd met behulp van verschillende statistische indices op verschillende ruimtelijke en temporele domeinen. Het resultaat laat zien dat de regenvalgegevens van CHIRPS beter waren dan alle andere producten op alle temporele en ruimtelijke schalen. Vervolgens kwamen schattingen van RFEv2-, 3B42v7- en PERSIANN-producten het dichtst in de buurt van metingen op regenmeters zowel voor de dagelijkse, maandelijkse en seizoensschalen als op punt- en ruimtelijke schalen. De prestaties van alle producten verbeterden naarmate de temporele schaal steeg tot maand en seizoen op alle ruimtelijke schalen. In vergelijking met de laaglanden steeg de Percentage Bias (PBias) in hoger gelegen gebieden met 35%, terwijl de correlatiecoëfficiënt (r) met 28% daalde. CHIRPS en 3B42v7 producten toonden de beste overeenkomst in bergachtige terreinen. CMorph en 3B42v7 overschatten de regenval consequent ten opzichte van alle regenmeters tijdens de pixel-tot-punt benadering en in laaglandgebieden tijdens de ruimtelijke gemiddelde regenvalvergelijking. De andere zes producten onderschatten de regenval op alle ruimtelijke schalen. Samenvattend, schattingen van regenval door CHIRPS komen het best overeen met grondwaarnemingen van regenval onder alle omstandigheden. CHIRPS werd daarom gebruikt voor validatie en invulling van ontbrekende grondobservaties van regenval en als input voor hydrologische modellering in deze studie.

De lange termijn trends en koppelingen tussen regenval en rivierdebiet werden geanalyseerd voor 21 regenval en 9 afvoer stations om de mogelijke drijvende krachten achter veranderende afvoervariabelen in het bekken te identificeren. Trendanalyse en detectie van veranderpunten van deze variabelen werden geanalyseerd met behulp van de Mann-Kendall- en Pettitt-tests. Fluctuaties in deze variabelen werden ook onderzocht met behulp van indicatoren van hydrologische verandering (IHA). Uit de analyse van de trends en het veranderpunten bleek dat 20 van de 21 geteste regenstations in de afgelopen 63 jaar geen statistisch significante veranderingen vertoonden. Daarentegen vertoonde de rivierdebiet zowel significante stijgende als dalende patronen. Zes van de negen afvoer-meetstations vertoonden een dalende trend in het droge (oktober tot februari), korte (maart tot mei), belangrijkste regenseizoenen (juni tot september) en jaarlijkse totalen. Slechts één van de negen meetstations kende tijdens het droge en korte regenseizoen een aanzienlijke toename van de rivierafvoer; dit kan worden toegeschreven aan de bouw, in 2009, van de Tekeze waterkrachtdam stroomopwaarts van dit station in 2009.

De afvoertrends en tijdstippen van veranderpunten bleken niet consistent te zijn tussen de stations. Veranderingen in de riverafvoer zonder significante verandering in neerslag suggereren dat andere factoren dan regenval die veranderingen sturen. Dit geeft aan dat de waargenomen veranderingen in het afvoerregime mogelijk kunnen worden toegeschreven aan de verandering in stroomgebiedskenmerken.

Daarom werden eerst de veranderingen in ruimte en tijd van landgebruik/bodembedekking (LULC) geanalyseerd. Daartoe werden de door de mens veroorzaakte landschapstransformaties in het stroomgebied van Geba (~5.000 km²), de bovenloop van het studiebekken, gedurende de laatste vier decennia (1972-2014) onderzocht. Satellietbeelden, Geografisch Informatie Systeem (GIS) en grondobservaties werden gebruikt om de LULC in het stroomgebied te classificeren en veranderingen in de tijd te detecteren. Bovendien identificeerde een waarschijnlijkheidsmatrix systematische overgangen tussen de verschillende LULC-categorieën en toonde aan dat meer dan 72% van het landschap in de afgelopen 43 jaar van categorie is veranderd.

De natuurlijke vegetatie daalde drastisch met de snelle expansie van agrarische en onbedekte gebieden gedurende de eerste twee decennia. De vegetatie begon zich echter te herstellen sinds de jaren 1990, toen enkele van de landbouw- en onbedekte gebieden in begroeide gebieden veranderden. Natuurlijke bosgebieden vertoonden tot 2001 een continu afnemend patroon, waarna die in de laatste periode (2001-2014) met 28% toenamen. De toename van vegetatie is het resultaat van intensieve programma's voor het beheer van stroomgebieden in de afgelopen twee decennia. Deze bevindingen waren belangrijk voor het verbeteren van ons begrip van de relatie tussen hydrologische processen en veranderingen in het milieu in het stroomgebied.

Deze relatie werd onderzocht met behulp van een geïntegreerde benadering bestaande uit hydrologische respons van LULC-veranderingen, het beoordelen van de verandering van de rivierafvoer met behulp van indicatoren van hydrologische wijziging (IHA) en het kwantificeren van de bijdrage van individuele LULC-typen aan de hydrologie met behulp van het Partial Least Square Regression-model (PLSR). Een ruimtelijk verdeeld hydrologisch model gebaseerd op het Wflow-PCRaster / Python-modelleerraamwerk werd ontwikkeld om de hydrologische processen van de eerder geïdentificeerde LULC-veranderingspatronen te simuleren. De resultaten tonen aan dat de uitbreiding van landbouw- en graasland ten koste van natuurlijke vegetatie de directe oppervlakkige afvoer met 77% deed toenemen en de droge seizoenafvoer met 30% verminderde in de jaren 90 ten opzichte van de jaren 70. Vanaf het einde van de jaren negentig begon de natuurlijke vegetatie echter te herstellen en de droge seizoenafvoer steeg met 16%, terwijl de oppervlakkige afvoer en jaarafvoer met respectievelijk 19% en 43% daalden.

Opvallende variaties in de veranderingen van de rivierafvoer werden opgemerkt in de Siluh, Illala en Genfel deelstroomgebieden, voornamelijk geassocieerd met de ongelijke ruimtelijke verdeling van landdegradatie en –rehabilitatie interventies. In de jaren 2010 stopte de stijging van de lage afvoer echter, hoogstwaarschijnlijk door een toename van wateronttrekkingen ten behoeve van irrigatie. Veranderingen in IHA parameters waren in overeenstemming met de waargenomen LULC-veranderingen. Uit de PLSR-analyse bleek dat de meeste LULC-typen een sterke associatie met alle hydrologische componenten vertoonden. Deze bevindingen tonen aan dat veranderingen in hydrologische omstandigheden inderdaad kunnen worden toegeschreven aan de waargenomen LULC-veranderingsdynamiek.

Bovendien werden de effecten van fysieke bodem- en waterconserveringsinterventies op de lage afvoer gekwantificeerd met behulp van een combinatie van gepaarde (controle en behandelde) en model-naar-model ("voor en na" interventies) vergelijkende benaderingen. De algehele impact van LULC-verandering kan de gekwantificeerde impact van dergelijke interventies niet uniek identificeren. Grootschalige implementatie van fysieke bodem- en waterconserveringsstructuren (SWC) kan de hydrologie van een stroomgebied wijzigen door de verdeling van de binnenkomende regenval op het landoppervlak zowel op positieve als op negatieve manieren te veranderen. Daarom is een wetenschappelijk begrip van de respons van lage afvoeren op SWC-interventies van cruciaal belang voor effectieve beleidsinterventies voor waterbeheer.

De resultaten toonden aan dat het volledig behandelde deelstroomgebied ($\sim$500 km^2) een aanzienlijke verandering heeft ondergaan na intensieve SWC-implementatie in grote delen ervan. Vergeleken met het controlebekken was de lage afvoer in het behandelde stroomgebied meer dan 30% groter terwijl de directe piekafvoer meer dan 120% lager was. Dit kan worden verklaard doordat een groot deel van de regenval in het behandelde stroomgebied infiltreert en grondwater oplaadt, wat later bijdraagt aan de basisafvoer tijdens de droge seizoenen. Het aandeel bodemopslag was meer dan het dubbele in vergelijking met het controlebekken als gevolg van de SWC-interventies die de infiltratiecapaciteit van het stroomgebied verbeterden. Hydrologische vergelijking in een enkel stroomgebied (model-tot-model) toonde ook een drastische vermindering van directe afvoer (> 84%) en een toename van de lage afvoer met meer dan 55% na de SWC interventies. Deze bevindingen werden bevestigd door de waargenomen veranderingen in het hydrologische regime met behulp van de IHA-methode. Hoewel de lage afvoer in het stroomgebied aanzienlijk toenam, daalde de totale afvoer aanzienlijk na grootschalige SWC-implementatie; dit wordt toegeschreven aan de toename van de opslag van bodemvocht en gewasverdamping, alsmede een toename van geïrrigeerde percelen en de daarmee geassocieerde water onttrekkingen uit de rivier. Dit heeft een negatieve invloed op de beschikbaarheid van blauw water voor benedenstroomse watergebruikers.

Concluderend heeft dit proefschrift aangetoond dat de lopende stroomgebiedsbeheersing activiteiten in het Upper Tekeze bekken de beschikbaarheid van water op verschillende ruimtelijke en temporele schalen op verschillende wijzen heeft beïnvloed. Significante veranderingen in de omvang van afvoercomponenten (bijv. jaartotalen, natte en droge seizoenafvoeren) werden op alle ruimtelijke schalen gevonden. De mate van verandering in de rivierafvoer van grotere stroomgebieden blijkt echter kleiner te zijn in vergelijking met dat van kleinere stroomgebieden. Dit typische schaaleffect wordt voornamelijk geassocieerd met de ongelijke ruimtelijke verdeling van beheers interventies in deelstroomgebieden.

Op basis van de resultaten van de lange-termijn trendanalyse van regenval en rivierafvoer, gedetailleerde analyse van veranderend landgebruik/bodembedekking (LULC) op de lange termijn, modellering van hydrologische respons op verandering van landgebruik en de kwantificering van hydrologische respons op SWC-interventies, heeft dit proefschrift het begrip verbeterd van hoe door de mens veroorzaakte milieuveranderingen van invloed zijn op hydrologische processen in het stroomgebied van de Upper Tekeze en de impact ervan gekwantificeerd. De gecombineerde analyse van neerslag-afvoermodellering, veranderingsindicatoren (IHA) en PLSR wordt aanbevolen om de impact van omgevingsverandering op de hydrologie van complexe stroomgebieden te evalueren. De IHA-methode is een robuust hulpmiddel om de omvang van veranderende rivierafvoeren te bepalen, terwijl met de PLSR-methode te identificeren is welke LULC verantwoordelijk is voor deze wijziging. De resultaten van deze studie kunnen het waterbeheer en het stroomgebiedsbeheer op verschillende schaalniveaus informeren, zowel stroomopwaarts als stroomafwaarts van het Upper Tekeze stroomgebied, op een manier dat het bekken zich op een duurzame manier kan ontwikkelen.

ABOUT THE AUTHOR

Tesfay Gebretsadkan Gebremicael was born in Adwa, Ethiopia, on September 13th, 1982. He received his BSc degree in Agricultural Engineering from Hawasa (the then Debub) University, Ethiopia, in July 2006. After a few months, he joined the Tigray Agricultural Research Institute (TARI) where he worked as an assistant researcher in water resources and irrigation engineering. He was involved in developing of proposals and implemented research activities related to irrigation water management, evalution of irrigation structures and sedimentation. From October 2009 to June 2011, he followed the MSc programme in Water Management, at UNESCO-IHE Institute for Water Education in Delft, The Netherlands. After obtaining his MSc degree, he returned to Ethiopia and continued to work with TARI as researcher in water resources and hydrology. From July 2011 to February 2015, he actively participated in designing and implementing large research projects which focused on hydrological, sediment and water resources modelling, irrigation water management, water harvesting structures and spatial analysis.

In April 2015, he received a scholarship from the Netherlands Fellowship Program (NFP) for his PhD studies at UNESCO-IHE. During his PhD research, he studied the hydrology of the Upper Tekeze basin at various spatio-scales using a combination of different methods. During his PhD research work, he specialized in catchment hydrology, water resources management, hydrological and river systems modelling and spatial and remote sensing analysis. Moreover, he has supervised four MSc students and served as a reviewer for several international peer-reviewed journals, including Journal of Hydrology, Hydrology and Earth System Sciences, Science of the Total Environment, Journal of Arid Environments, and Water Resources Research. He has presented his work at various national, regional and international conferences and has published several articles in international journals and conferences.

Journal publications

Gebremicael, T.G., Mohamed, Y.A., Van der Zaag, P., (2019a). Attributing the hydrological impact of different land use types and their long-term dynamics through combining parsimonious hydrological modelling, alteration analysis and PLSR analysis. *Science of the Total Environment*, 660, 1155-1167.

Gebremicael, T.G., Mohamed, Y.A., van der Zaag, P., Gebremedhin, A., Gebremeskel, G., Yazew, E., Kifle, M., (2019b). Evaluation of multiple satellite rainfall products over the rugged topography of the Tekeze-Atbara basin in Ethiopia, *International Journal of Remote Sensing*, 1-20.

Gebremicael, T. G. Mohamed, Y., van der Zaag, P., Yazew. E., (2018). Quantifying longitudinal land use change from land degradation to rehabilitation in the headwaters of Tekeze-Atbara basin, Ethiopia, *Science of the Total Environment*, 622-623, 1581-1589.

Gebremicael, T. G. Mohamed, Y., van der Zaag, P., Yazew. E., (2017). Temporal and spatial changes of rainfall and streamflow in the Upper Tekezē–Atbara river basin, Ethiopia. *Hydrology and Earth System Sciences*, 21(4):2127–2142.

Mesfin, H, Gebremedhin, A, **Gebremicael, T.G.**, (2018). Hydrological response under climate change scenario using SWAT Model: The case of Ilala watershed, Northern Ethiopia, *Modeling Earth Systems and Environment*, 4(1), 437-447.

Gebremedhin, A. Mefin., H., Abrah, A., Abraha, G., Misgina, S., **Gebremicael, T.G.** (2018). Impact of Climate Change on net Irrigation Water Requirement of major crops in the semi-arid regions of Northern Ethiopia, *Journal of the Drylands, 8(1), 729-740*

Kifle, M., **Gebremicael, T. G.**, Girmay, A., and Gebremedihin, T., (2017). Effect of surge flow and alternate irrigation on the irrigation efficiency and water productivity of onion in the semi-arid areas of North Ethiopia, *Agricultural Water Management*, 187, 69-76.

Kifle, M., Dimtsu, G. Y., **Gebremicael, T.G.** (2018). Evaluation of the effect of mouldboard plow and tied-ridger on wheat productivity in Atsibi and Ganta Afeshum districts, Ethiopia. *Advances in Agricultural Science*, 6(3), 17-24.

Gebremeskel, G., **Gebremicael, T. G.**, and Girmay, A., (2017). Economic and environmental rehabilitation through soil and water conservation, the case of Tigray in northern Ethiopia, *Journal of Arid Environments*, 151, 113-124.

Gebremeskel, G., **Gebremicael, T.**, Hagos, H., Gebremedhin, T., and Kifle, M., (2017). Farmers' perception towards the challenges and determinant factors in the adoption of drip irrigation in the semi-arid areas of Tigray, Ethiopia, *Sustainable Water Resources Management*, 1-11,

Kifle, M., and **Gebremicael, T. G.**, (2016). Yield and water use efficiency of furrow irrigated potato under regulated deficit irrigation, Atsibi-Wemberta, North Ethiopia, *Agricultural Water Management*, 170, 133-139.

Gebremicael T. G, Mohamed, Y.A., Betrie, G.D., van der Zaag, P., Teferi, E., (2013). Trend analysis of runoff and sediment fluxes in the Upper Blue Nile basin: A combined analysis of statistical tests, physically-based models and land use maps. *Journal of Hydrology*. 82, 57–68.

Gebremicael, T.G., Mohamed, Y.A., Van der Zaag, P., (2019c). Change in low flows due to catchment management dynamics - a comparative catchment modelling approach in the headwaters of Tekeze-Atbara basin. Hydrological Processes (Under review).

Gebremeskel, G., **Gebremicael, T. G.**, Mulubrehan Kifle, Teferi Gebremedhin. (2019). Effects of irrigation scheduling and irrigation methods on onion productivity in semi-arid irrigation schemes in northern Ethiopia. Agricultural Water Management (under review).

Selected conference proceedings and presentations

Gebremicael, T.G., Mohamed, Y.A., P. van der Zaag. (2019). Hydrological response to land management dynamics in the headwaters of Tekeze-Atbara basin. Application of parsimonious hydrological modelling, alteration and PLSR analysis, *Presented at Upper Tekeze water related studies conference*, March 24-26, 2019, Khartoum, Sudan.

Gebremicael, T.G., Mohamed, Y.A., P. van der Zaag., (2018). Hydrological response to land degradation and rehabilitation in the important tributary of the Nile River basin using Wflow-

PCRaster/python modelling framework. *Presented at AGU 2018 Fall meeting*, December, 10-14, 2018, Washington Dc, USA.

Gebremicael, T.G., Mohamed, Y.A., P. van der Zaag, Gebremedhin, A., Gebremeskel, G., Yazew, E., Kifle, M., (2018). Validation of multiple satellite rainfall products over the complex topography of Upper Tekeze tributary of the Nile

Gebremicael, T.G.,Mohamed, Y., P. van der zaag, (2017). Hydro-climatic variabilities in the Upper Tekeze headwaters, *Presented at the Inception workshop for river simulation for improved transboundary water managment in the Nile*, May, 22-23, 2018, Gedarif, Sudan.

Gebremicael, T.G., Mohamed, Y., van der zaag, P., Yazew. E., (2016). Drivers of the spatio-temporal variabilities of hydrological flows in the Tekeze river basin, Ethiopia. *International conference on "Sustainable Land and Watershed Management (SLWM3) for building resilience against climate change"*, 28-30/11/2016, Mekelle, Ethiopia.

Gebremicael, T. Mohamed, Y., van der zaag, P., Yazew. E., (2016). Modelling the impact of catchment management dynamics on the hydrological process in the headwaters of Upper Tekeze River Basin. *PhD Symposium on Integrating Research Water Sector*, UNESCO-IHE, 28-29, 2016, Delft, The Netherlands.

Mulubrehan Kifle, **Gebremicael, T.G.**, Eyob Kahisay, (2014). Irrigation efficiency andwaterproductivity of furrow irrigation in dry land areas of Tigray, Ethiopia. *International conference on sustainable land and watershed management*, 26th-27th may 2014, Mekelle University, Mekelle, Ethiopia.

Gebremicael, T.G., Mulubrhan Kifle and Awetahnge Nigus, (2013). The Effect of existing Soil and Water conservation practices on natural resources improvement of Tigray Region Using remote sensing and SWAT model. *Regional conference on Soil and water research*, 29-30, June 2013, Mekelle, Ethiopia.

Mulubrehan Kifle, **Gebremicael T.G.**,Tesfay Asgedom and Teferi Gebremedhin., (2013). Comparisons of furrow and drip irrigated crops under deficit irrigation. *Regional conference on Soil and water research*, 29-30, June 2013, Mekelle, Ethiopia.

Gebremicael, T.G., Mohamed, Y.A., Betrie, G.D., van der Zaag, P., Teferi, E., 2013. Trend analysis of runoff and sediment fluxes in the Upper Blue Nile basin: A combined analysis of statistical tests, physically-based models and land use maps. International conference on: *New Nile Perspectives Scientific advances in the Eastern Nile Basin*, 6th-8th may 2013, Khartoum, Sudan.

Gebremicael, T.G., Yasir Mohamed, Getnet Dubale, Pieter Van der Zaag and Ermias Teferi, 2012. Trend Analysis of Runoff and Sediment fluxes using Statistical and physically based models: Upper Blue Nile Basin. *Proceedings of National conference on Science, Technology and innovation for prosperity of Ethiopia (NCSTI-2012)*, 16-16, may 2012, Institute of Technology, Bahirdar University, Ethiopia.

Netherlands Research School for the
Socio-Economic and Natural Sciences of the Environment

D I P L O M A

For specialised PhD training

The Netherlands Research School for the
Socio-Economic and Natural Sciences of the Environment
(SENSE) declares that

Tesfay Gebretsadkan
Gebremicael

born on 13 September 1982 in Adwa, Ethiopia

has successfully fulfilled all requirements of the
Educational Programme of SENSE.

Delft, 8 October 2019

The Chairman of the SENSE board

Prof. dr. Martin Wassen

the SENSE Director of Education

Dr. Ad van Dommelen

The SENSE Research School declares that Tesfay Gebretsadkan Gebremicael has successfully
fulfilled all requirements of the Educational PhD Programme of SENSE with a
work load of 58.7 EC, including the following activities:

SENSE PhD Courses

- Environmental research in context (2015)
- Research in context activity: 'Arid African Alluvial Aquifers): Kick off workshop on Arid
  African Alluvial Aquifers Labs Securing Water for Development (Hawzen, Tigray,
  Ethiopia, 6-7 march, 2017) and International conference on Arid African Alluvial Aquifers
  Labs Securing Water for Development (Wukro, Tigray, Ethiopia, 13-16 March 2017)'

Selection of other PhD and Advanced MSc Courses

- Transferable and competency course, TU Delft (2015)
- The Informed Researcher: management of information and Data Skills, TU Delft (2015)
- Scientific writing and presenting courses: 'Creative tools for Scientific writing', 'Analytic
  storyline in writing', and 'The art of presenting science", TU Delft (2016- 2017)
- Career development – preparing for your next step in academia (2018)

Management and Didactic Skills Training

- Teaching of research database and reference management skills (Endnote and
  Mandeley) to researchers of Tigray Agricultural Research Institute
- Supervising four MSc students with thesis (2017-2018)
- Contribution to development of project proposal entitled " Integrated Research,
  development and capacity building of Hamedo irrigation project" in Tigray, Ethiopia. For
  Tigray regional government
- Water accounting training for trainers , IHE Delft (2018)
- PhD association board member (2016-2018)

Oral Presentations

- *Modelling the impact of catchment management dynamics on the hydrological process
  in the headwaters of Tekeze-Atbara River Basin, Nile River.* UNESCO-IHE annual PhD
  symposium, 28-29 September 2015, Delft, The Netherlands
- *Drivers of the spatio-temporal variabilities of hydrological flows in the Tekeze river basin,
  Ethiopia.* International conference International conference on "Sustainable Land and
  Watershed Management (SLWM3) for building resilience against climate change, 28-20
  November 2016, Mekelle, Ethiopia
- Hydrological response to land degradation and rehabilitation. AGU 2018 fall meeting,
  10-14 December 2018, Washington DC, USA